CHINA IN 2020

THE THORNTON CENTER CHINESE THINKERS SERIES

The John L. Thornton China Center at Brookings develops timely, independent analysis and policy recommendations to help U.S. and Chinese leaders address key long-standing challenges, both in terms of Sino-U.S. relations and China's internal development. As part of this effort, the Thornton Center Chinese Thinkers Series aims to shed light on the ongoing scholarly and policy debates in China.

China's momentous socioeconomic transformation has not taken place in an intellectual vacuum. Chinese scholars have actively engaged in fervent discussions about the country's future trajectory and its ever-growing integration with the world. This series introduces some of the most influential recent works by prominent thinkers from the People's Republic of China to English language readers. Each volume, translated from the original Chinese, contains writings by a leading scholar in a particular academic field (for example, political science, economics, law, or sociology). This series offers a much-needed intellectual forum promoting international dialogue on various issues that confront China and the world.

Also in this series:

Yu Keping, *Democracy Is a Good Thing: Essays on Politics, Society, and Culture in Contemporary China*, 2009

CHINA IN 2020

A New Type of Superpower

HU ANGANG

BROOKINGS INSTITUTION PRESS
Washington, D.C.

Library of Congress Cataloging-in-Publication data

Hu, Angang.
   China in 2020 : a new type of superpower / Hu Angang.
      p.   cm.
   Includes bibliographical references and index.
   Summary: "Explains how China, as the world's largest emerging market, will impact
global economic growth, FDI flows, energy consumption, climate change, and other
arenas, with a proposed strategic framework that would guide the country's rise while
maximizing positive impacts and minimizing any negative externalities"—Provided by
publisher.
   ISBN 978-0-8157-0478-2 (cloth : alk. paper)
   1. China—Economic conditions—2000– 2. China—Social conditions—2000–
3. Health status indicators—China. 4. Environmental indicators—China. 5. Science
indicators—China. 6. Technology indicators—China. I. Title.
   HC427.95.H817 2011
   303.4951—dc22   2011001325

9 8 7 6 5 4 3 2 1

Printed on acid-free paper

Typeset in Adobe Garamond

Composition by Cynthia Stock
Silver Spring, Maryland

Printed by R. R. Donnelley
Harrisonburg, Virginia

CONTENTS

Foreword

JOHN L. THORNTON

We began the Chinese Thinkers Series to expose those outside China, especially the English-language audience, to the depth, complexity, and richness of the intellectual discussion occurring within China today. The first volume of the series, *Democracy Is a Good Thing*, introduced the works of the political theorist Yu Keping. The present book collects the forceful ideas of my fellow Tsinghua University faculty member Hu Angang, whose influence on Chinese thinking in the economic sphere matches that of Yu's in politics and governance.

Professor Hu has advised the Chinese leadership directly and is frequently consulted by a wide range of senior officials at both the central and provincial levels. The government has drawn on his expertise in drafting the five-year development plans that serve as the main road map for economic policy. Intense, analytical, and emotional, Professor Hu gives the impression of keeping a tight lid on a potential overflow of ideas and energy. He is an inexhaustible seeker of data and insight, always oriented toward the future. No Chinese thinker has a better record of predicting the pace and direction of the country's development. When he is certain, he is bold. He is of the school that would prefer to be generally right than precisely wrong. He is probably China's most versatile and pragmatic economist.

Some eight years ago, I invited Professor Hu to co-teach a session of my Global Leadership seminar at Tsinghua University's School of Economics and Management. He believes, as I do, that young Chinese—the country's future leaders—have a special responsibility to appreciate the context of China's rapidly changing place in the world, develop a global view, and orient their lives to be a force for common good in China and abroad.

In the class we taught together, Professor Hu explained the compelling concept of Green GDP, which factors in the cost of environmental damage in calculating growth rates. The possibility of adopting Green GDP as the national measure of economic growth was actively discussed and seriously considered within the Chinese government. Though the initiative has been tabled for now, Hu's persistent campaign to explain his concept and the reasoning behind it has led to a meaningful shift in understanding and emphasis among Chinese leaders about the right kind of growth for the country. As a matter of long-term strategic national planning, the Chinese government today knows that it must build a "new model" of development that is cleaner and more energy efficient if it is to continue to move China forward for the benefit of its people. In the pursuit of this goal, the recently unveiled Twelfth Five-Year Plan put forward green development as a new principle informing China's medium-term development.

Hu notes that nearly every prediction of China's growth since the 1980s, including that of the Chinese government itself, has consistently and significantly underestimated the pace of actual development that followed. (For example, as Hu notes, "As of the year 2000 China's total GDP was 78.5 percent larger than predicted by the World Bank.") Now at a time when skeptics are once again asking whether China is experiencing a dangerous and unsustainable economic bubble, Hu rejects these "fallacies" and argues why he believes China's growth is real and well-grounded, even while acknowledging the country's many remaining challenges. Hu has no doubt that China is destined to be a superpower.

The question, of course, is what kind of superpower it will be. Hu's view is that, for historical and cultural reasons, China will strive to become a "mature, responsible, and attractive superpower." Hu sees the world entering a multipolar era when American supremacy is replaced by the interaction of a number of great powers, including China. Unlike previous eras, however, he believes the new order will be "defined less by competition than cooperation." China will find it in its interest to be cautious, not radical, and reject hegemony in favor of continued peaceful development. "China does not want to replace the United States and become the sole leader of the world," Hu argues in the book. His analysis, of course, is open to debate by the international community as it is in China.

Professor Hu is a forceful proponent of his strong views, advocating an optimistic and constructive vision for his country to both decisionmakers and the Chinese public. His ideas have measurable impact. That is why it is important for those of us who are not Chinese, yet seek to understand the country and where it is likely headed, to pay heed to this distinctive thinker and his provocative writings. If this book helps accomplish that, it will be a modest but important contribution to ensuring the human condition improves in the twenty-first century.

ACKNOWLEDGMENTS

As China is rising on the international stage, China studies have become one of the most popular fields in academia. In the PRC, a dozen China research centers have been set up at internationally renowned academic institutions and think tanks, many scholars have worked on interdisciplinary China studies, and publishers in China and overseas have produced books on a range of topics within the fields of Chinese economics and sociopolitical development. As an observer, researcher, participant, and facilitator, I have published several books focusing on China's future trajectory, including *China in 2020: Building a Well-Off Society* (2007) and *China Towards 2015* (2010). These books provide systematic knowledge, specific data, and scholarly analysis for Chinese readers to better understand contemporary China and propose development strategies and policy recommendations for the future.

On the invitation of the John L. Thornton China Center at the Brookings Institution, I wrote this book from an insider's perspective: employing general theories and comparable statistics to interpret processes, features, and influences on China's road to superpower status. There is no question in my mind that China will become a superpower, despite many unprecedented challenges to China's development. The key issue is what type of superpower. The subtitle of the book, *A New Type of Superpower,* indicates the uniqueness of China's road to modernization when compared to the United Kingdom, the United States, and the former Soviet Union. I hope this book can provide a distinct Chinese perspective and reasonable explanations, as well as new information that will give English readers a more objective and comprehensive view of China's rise.

I would like to first express my gratitude to John L. Thornton, chairman of the board of the Brookings Institution and professor at the School of Economics and Management at Tsinghua University. The book was begun with his warm invitation and was completed largely thanks to his constant encouragement and enduring support. During my conversations with Mr. Thornton over the years I have learned much from his sagacity. He also took time out of his very busy schedule to write the insightful foreword, for which I am very grateful. I would like to also acknowledge Woo Lee for his help and input on the project.

I want to thank Dr. Cheng Li, director of research at the John L. Thornton China Center at the Brookings Institution. He has offered his great help during every step of the book project, from topic selection to many rounds of revision. With his personal warmth and professional high standards, he led many of his wonderful assistants at the China Center, including Eve Cary, Jordan Lee, Robert O'Brien, and Paul Wozniak, in polishing the manuscript in the process of publication. Many of the book's improvements should be attributed to their diligent work. In particular, I would like to thank Dr. Li for his excellent introductory chapter. I would be remiss not to also thank the leadership at Brookings: Strobe Talbott, president, Martin Indyk, vice president and director of Foreign Policy, Ted Piccone, deputy director of Foreign Policy, and Kenneth Lieberthal, director of the John L. Thornton China Center, for their continual guidance and support.

I would also like to express my appreciation for the work of Janet Walker, managing editor of the Brookings Institution Press, and Diane Hammond, editor of the book, for their very efficient and effective editing. Brookings Institution Press director Bob Faherty, marketing director Chris Kelaher, production manager Larry Converse, and art coordinator Susan Woollen also provided immense support throughout the publication process of this volume. Thanks also go to an anonymous reviewer of this volume for his or her time and for excellent and helpful comments.

Finally, I would like to thank Mr. Shaoqing Jin and Dr. Xing Wei for their help with translation, research assistance, and proofreading. I, of course, take full responsibility for the book's accuracy. The arguments presented here represent my beliefs and analysis, and they do not necessarily reflect those of the Brookings Institution or any associated scholars.

Currently, the Center for China Studies at Tsinghua University is carrying out research on a new project, titled *China in 2030,* which attempts to forecast the nature of China's long-term development and provide strategic thinking for Chinese policymakers and knowledge for the public. I hope that I can (in a humble way) continue to help international readers develop a more comprehensive understanding of a rapidly changing China.

HU ANGANG
Tsinghua University, Beijing
March 2011

INTRODUCTION

A Champion for Chinese Optimism and Exceptionalism

CHENG LI

> To see ourselves as others see us is a rare and valuable gift, without a doubt. But in international relations what is still rarer and far more useful is to see others as they see themselves.
>
> JACQUES BARZUN

China perplexes the world. The country's rapid rise to global economic power poses an important set of questions regarding how one should perceive the transformation of the international system in light of this epochal change:

—Is China on track to become a new superpower? If so, how will this transform the global economic and political landscape?

—Will this ongoing power shift be comparable in scale to the rise of Europe in the seventeenth century or the rise of America in the late nineteenth and early twentieth centuries?

—Will the world witness increasingly intense competition between the United States, the existing superpower, and China, an emerging superpower? Could it even lead to the outbreak of what international relations scholars call a hegemonic war?[1]

—Might a new cold war take shape as China, a Leninist one-party state, comes to rival the West in the decades ahead? Will China present a military

The author thanks Sally Carman, Eve Cary, Sean Chen, and Jordan Lee for their very helpful comments on an early version of this introductory chapter.

and ideological challenge to the West, as the Soviet Union did during most of the latter half of the twentieth century?

—Conversely, should the rise of the world's most populous country be seen as an auspicious development, able to fuel global economic growth and contribute to a more balanced and stable world order?

At this point there are no definitive answers to these questions, and increasingly sophisticated assessments of China's quest for superpower status emerge over time.[2] This type of analysis is also difficult, as the real and substantive impact of China's rise on the international system will depend on many factors. To a large extent, China's own economic and political trajectories—as well as the country's popular aspirations and demographic constraints—are the factors that will determine the role that China adopts. The momentous socioeconomic transformation propelling these changes has not occurred in an intellectual vacuum. In fact, over the past decade strategic thinkers and public intellectuals in China have engaged in fervent discussions of the nature of China's ever-increasing integration into the world and the country's road ahead.

Unfortunately, English-language studies of present-day China have not adequately informed a Western audience of the dynamism of the debates within China and the diversity of views concerning its own future.[3] In such a rapidly changing and complex world, it would be enormously valuable for the decisionmakers and analysts in the West to broaden their perspective and "see others as they see themselves," as the distinguished historian Jacques Barzun wisely suggests.[4] The international community's discourse on the implications of a rising China will increase its sophistication if it pays greater heed to how Chinese intellectuals perceive and debate the responsibilities that China may assume in the future. In particular, the American China-watching community would be much better informed if it were more familiar with the contemporary strategic discourse of the People's Republic of China (PRC).

The Influence of Hu Angang in the Chinese Discourse on China's Rise

Arguably no scholar in the PRC has been more visionary in forecasting China's ascent to superpower status, more articulate in addressing the daunting demographic challenges that the country faces, or more prolific in proposing policy initiatives designed to advance an innovative and sustainable economic development strategy than Hu Angang, the author of this volume. His strong influence on the Chinese intellectual and policy debates concerning the country's future is especially evident in three respects:

—For over two decades Hu has been forecasting China's socioeconomic and demographic development and has also established a popular index of comprehensive national power. In his 1991 book, *China: Toward the 21st Century,* Hu accurately forecast that China would emerge as a global economic giant sometime in the first or second decade of the twenty-first century, surpassing France, England, and Germany.[5] He was officially involved in drafting the Chinese government's five-year plans, which outline the government's key development goals.

—As a scholar well known for his concerted effort to break down strictly defined academic boundaries, Hu's remarkably broad research interests include demography, ecology, education, public health, environmental protection, anticorruption, and international relations. As early as 1988 Hu and two of his colleagues at the Chinese Academy of Sciences (CAS) showed great foresight in a well-documented, extensive report arguing that the "ecological deficit [*shengtai chizi*] will be the greatest liability for China's development in the 21st century."[6] Hu was among the first Chinese scholars to call for the need to measure "green GDP" in regional development.[7]

—Hu has not only authored or coauthored nearly sixty books and edited volumes (see the list of Hu's writings at the end of this book), he has also written and edited more than 900 reports on China studies (*guoqing baogao*) through the Center for China Studies, an influential think tank that is now affiliated with both CAS and Tsinghua University. Hu founded the center in 2000 and has served as its director ever since. These reports have been primarily circulated among ministerial and provincial leaders and higher authorities.[8] For example, of the thirty-seven reports that were submitted to the State Council during 2007–10, senior leaders of the State Council commented on these reports thirty-nine times.[9]

Hu Angang's active participation in the strategic thinking behind China's development over the past two decades reflects the growing role of public intellectuals and think tanks in the formation of the country's domestic and foreign policies.[10] Economic globalization and China's increasing importance in the world economy, and the technical and specialized knowledge that decisionmakers thereby require, have understandably resulted in more substantive input from economists and other specialists. Meanwhile, China's booming publishing industry and expanding mass media (both old and new) provide unprecedented opportunities for scholars like Hu Angang to articulate their views, exert their influence, and shape public opinion. While Hu appears to enjoy his role as an informal adviser to senior leaders in the Chinese government, he has maintained his primary role as an independent scholar. Indeed, from time to time Hu has criticized government policies and voiced concerns about possible policy pitfalls or crises in the making. This combination of close involvement and impartial detachment has afforded Hu Angang a vantage point from which he can exert some influence in decisionmaking circles, while his refusal to simply toe the party line has helped him establish credibility in the eyes of the Chinese public.

This book, *China in 2020: A New Type of Superpower,* is partly based on Hu Angang's 2007 Chinese book *China in 2020: Building a Well-Off Society.*[11] In this English-language volume, the author substantially expands the focus and content of the previous (Chinese) book. It covers many broad areas of China's rise on the world stage, often from a cross-country comparative perspective. In addition to the author's assessment of China's economic transformation, the book examines other important subjects, such as China's demographic trends, public health, education and human resources, science and technology, and approach to climate change, many of which are the focus of Hu's more recent Chinese publications. This volume's rich empirical data, multidisciplinary nature, explication of indigenous Chinese concepts, and thought-provoking arguments concerning China's rise make it invaluable for understanding China's role in today's world.

The Themes of the Book: Two Parallel Arguments

Two important themes that permeate most of Hu Angang's writings emerge in this volume: Chinese optimism and Chinese exceptionalism. Hu Angang has been consistently optimistic about China's socioeconomic transformation and its historic reemergence in the late twentieth and early twenty-first centuries, even during periods when the country was beset with serious challenges, such as the 1989 Tiananmen incident, the 1997 Asian financial crisis, the 2003 SARS epidemic, the 2008 Sichuan earthquakes, and the 2008 global economic meltdown. This does not mean that he has overlooked the many daunting challenges that China faces. On the contrary, over the past two decades Hu has often been ahead of other Chinese intellectuals in calling attention to such challenges as economic disparity, environmental degradation, energy inefficiency, public health crises, official corruption, and the loss of state assets. While many Chinese leaders and scholars have been encouraged by the fact that China is now the world's largest exporter and second-largest economy, Hu has eschewed triumphalism and chosen to remind the public that China has also become the world's largest carbon emitter and second-largest consumer of energy.[12]

While noting these serious problems, Hu still holds the optimistic view that China will continue its high-speed economic growth in the next decade and beyond due to a combination of factors. These include the country's solid industrial foundation, newly built world-class infrastructure, high rates of investment savings and foreign investment, large domestic market, human resource advantages, and last, but certainly not least, the country's commitment to transitioning toward a domestic, demand-driven, and environmentally friendly mode of economic growth. In this volume Hu argues that by 2020 China will likely not only surpass the United States as the largest economy in the world but also (because of its accomplishments in education, innovation, and clean energy) emerge as a "mature, responsible, and attractive superpower."

As for the second major theme, Hu acknowledges that the prevailing wisdom in Western international relationships scholarship holds that an emerging superpower will destabilize the existing international system due to zero-sum competition with the existing superpower over spheres of influence, natural resources, market access, and military superiority. But Hu believes that China's rise to superpower status will be an exception to the rule. In his words, China will constitute a "new type of superpower." He observes that in an increasingly interdependent world China has neither the resources nor the intention "to replace the United States and become the sole leader in the world. Rather, China needs to cooperate with the United States in order to cope with global challenges in economics, politics, energy, and the environment."

Of course, foreign analysts of China will not be so naïve as to take Hu's arguments at face value. Both Chinese optimism and Chinese exceptionalism will, and should be, subjected to continual scrutiny as Chinese decisionmakers adjust to constantly changing domestic and international circumstances. Many of Hu's propositions reflect an ideal rather than a reality, and they are properly understood to be well-intended aspirations and promises rather than predetermined conditions or inevitable prospects. Even if Hu genuinely believes—and I think he does—that China should neither pursue a strategy of replacing the United States as the sole superpower in the world nor adopt belligerent and bullying policies toward its neighboring countries, he cannot offer assurance that this will be the case. The logic and nature of superpower competition, so some international relations scholars argue, is likely to encourage the emerging superpower to behave in a more aggressive and hostile way toward the existing superpower.[13]

Foreign critics may also reasonably wonder how representative Hu's views are of China's mainstream intellectual and policy communities. In fact both

Chinese optimism and Chinese exceptionalism have a large number of domestic critics. It is important to note that Chinese assessments of the PRC's quest for superpower status and its implication for peace and prosperity in the world are as diverse and controversial in China as elsewhere. But foreign decisionmakers and analysts should not be too quick to discount the significance of Hu's prognostications. In a way, Hu actually proposes a comprehensive strategic framework for Chinese decisionmakers to guide the next stage of China's rise, seeking to maximize the country's positive impact on the world and minimize the negative effects of its meteoric development. The stakes of a conflict between superpowers are too high, and policymakers and public intellectuals in the United States, China, and elsewhere should explore all options and scenarios. Therefore, it is essential for foreign analysts to acquire a more nuanced and accurate understanding of this influential Chinese thinker, including his life experiences, professional background, scholarly pursuits, and overall worldview. Meanwhile, Hu's arguments should also be evaluated within the broader context of strategic thinking in the PRC, especially compared with the way other public intellectuals in the country view Chinese optimism and exceptionalism.

Hu Angang: From "Sent-Down Youth" to High-Profile Economist

Hu Angang was born to a family of intellectuals in Anshan City, Liaoning Province, in 1953. During the 1950s Anshan City was home to China's largest steel and iron factory and was considered the "capital of the steel industry of the PRC." The term Anshan steel (*angang*) became part of Hu's name, reflecting the socioeconomic and political environment in which he was born. That was, of course, an era in which the country was fanatically obsessed with the development of the steel industry. Hu's parents, natives of Zhejiang Province, were graduates of Shanghai's Jiaotong University, one of the top engineering schools in the country, and they worked in this northeastern industrial city after graduation. Hu's parents were once awarded the title National Model Worker for their devotion and contribution to the country's rapid drive for industrialization during the Mao era.

Hu Angang has three siblings (all brothers), all of whom attended college after the Cultural Revolution, obtained advanced academic degrees, and studied in North America (the United States or Canada). Hu Angang's wife, Zhao Yining, is the chief reporter for China's liberal newspaper *21st Century Business Herald* and a public intellectual in her own right. Her interviews with influential leaders in China and abroad, such as U.S. Secretary of the Treasury

Henry Paulson, have often spurred lively intellectual and policy discussions in the country. She is also the author of the best-selling book *Grand Games: When the Chinese Dragon Faces the American Eagle*.[14]

Hu Angang grew up during the Cultural Revolution, the era in which China's educational system—including elementary schools, middle schools, and colleges—was largely paralyzed. Chinese students were generally engaged in political campaigns and ideological indoctrination rather than academic work. Although catastrophic for the entire nation, the Cultural Revolution affected Hu Angang's age cohort the most. They were in elementary and middle school when the revolution began and suffered turbulent changes and extraordinary hardships during their adolescence. In 1969, at the age of sixteen, without finishing his middle school education in Beijing, Hu was sent to a collective farm in Beidahuang, the desert area of the Nen River Valley of China's northeast Heilongjiang Province (also known as the Great Northern Wilderness). Hu worked in this extremely arduous environment as a young farmer, or what the Chinese call a "sent-down youth" (*xiaxiang zhiqing*), for seven years. When the Cultural Revolution ended in 1976, Hu moved to Hebei Province where he served as a manual laborer on a geological team, continuing to work in the rural environment.

All of these arduous and humbling experiences, Hu says, actually helped him cultivate valuable traits such as diligence, endurance, adaptability, and humility.[15] The hardships in the countryside were so extreme that he not only remembers rural China but also developed a strong interest in economic inequalities, the regional and urban-rural gap, resource scarcity, and grain safety—topics that he would later focus on in great depth in his academic career. Hu's experiences led him to believe that "one who has no knowledge of rural China does not know about China; one who does not understand China's poverty-stricken regions does not have a real understanding of China."[16]

As for his undergraduate education, Hu Angang belonged to the famous class of 1982. This class of students, ranging in age from the late teens to the early thirties, passed the national entrance exams in late 1977 and early 1978 as a result of Deng Xiaoping's policy initiatives to select students by their academic credentials rather than political backgrounds. This group entered college in March and October of 1978, in two clusters, and graduated in 1982. The ratio of those who took the exam and those who were admitted to this class was twenty-nine to one, compared with a ratio of two to one in 2007.[17] This famous class was extraordinary not only for having passed the most competitive college entrance exams in the PRC's history but also for yielding many talented leaders in all walks of life.

After receiving a bachelor's degree in metallurgy from the Tangshan Institute of Technology, he earned a master's degree in metal pressure processing at the Beijing Institute of Steel and Iron (now the University of Science and Technology Beijing) in 1985 and a doctoral degree in engineering from the Institute of Automation of the CAS in 1988. Hu's undergraduate and graduate education seems to reflect the influence of his native city and his parents' professional careers in the steel industry. During his Ph.D. studies, however, Hu Angang became interested in researching China's socioeconomic development, a relatively new field in the country at that time. He participated in two research projects in 1985, which helped determine his professional career. The first was a feasibility study on the establishment of the State Information Center, and the other was a study on the use of a mathematical model to forecast and make strategic assessments of China's socioeconomic conditions in 2000.

According to Hu, two other events also strongly influenced his decision to devote his career to studying economics and demography.[18] One was the publication of a World Bank report in 1985, *China: Long-Term Development, Issues, and Options,* which offered comprehensive empirical information and laid out analytical paradigms describing the possible developmental paths and choices for China in 2000.[19] The other was the release of the Brundtland Commission's report, *Our Common Future,* in 1987.[20] This landmark report highlights the importance of sustainable development in the world and new international thinking and cooperation to meet new global challenges. In his Ph.D. program Hu also studied under Ma Shijun, who represented China in signing the Brundtland Declaration. Not surprisingly, Hu Angang frequently cites these two documents in this volume.

In 1988 Hu completed his doctoral dissertation, "Population and Development in China: A Systemic Analysis and Policy Measures," under the supervision of He Shanyu (an automation expert), Ma Bin (an economist), Zheng Yingping (a game theory expert), and Ma Zhengwu (a computer simulation expert). The dissertation focuses on the demographic impetus and constraints on China's economic development. In Hu's view China's large population and its characteristics constitute the most important factor for forecasting the country's future. After finishing his Ph.D., Hu began work at the Research Center for Eco-Environmental Sciences of the CAS. He played an instrumental role in the establishment of the national economic database of the center. Hu also greatly contributed to the national comprehensive power index, to which he often refers in this volume.

As both a prolific writer for scholarly publications and a frequent commentator in the Chinese media, Hu soon emerged as one of the highest profile public intellectuals in the country. In 1991 he was awarded the epithet, "PRC-trained Ph.D. with outstanding contributions," jointly by the State Education Commission and the State Council's Academic Degrees Committee. Hu was invited to join the Chinese Economists 50 Forum (*Zhongguo jingji wushiren luntan*), which includes the country's fifty most prominent economists (such as Lin Yifu, then Peking University professor and now senior vice president and chief economist of the World Bank, and professor Wu Jinglian, China Europe International Business School) and government technocrats (such as Zhou Xiaochuan, the governor of People's Bank, and Xiao Jie, director of the State Taxation Bureau). In 2000 Hu won a Sun Yefang Award in Economic Research, one of China's most prestigious prizes for scholarly publications in economics. Hu was also named a few times by the Chinese media as one of the top ten economists in the country.

Hu has earned these awards and public recognition largely because he is not an ivory tower, academic economist, interested exclusively in economic theories and mathematical modeling. Instead he is actively engaged in important public policy debates. For example, he has made frequent appeals in recent years—in both official briefings and media interviews—for China to promise to curb carbon emissions with a clearly outlined timetable.[21] In his view, this should be done not only to allay international pressure but also to transform the country's economic development model. Commenting on Deng Xiaoping's famous motto, "It doesn't matter if a cat is white or black, as long as it catches mice," Hu argues that for China to pursue sustainable development at home and be a responsible stakeholder in the twenty-first century, it must transform from "a black cat" to "a green cat."[22]

Hu Angang's Experience Abroad and Nationalistic Sentiments

Hu Angang spent much of the 1990s abroad, serving as a postdoctoral fellow at Yale University from 1991 to 1992, a visiting professor at Murray State University in Kentucky in 1993, a research fellow at the School of Arts and Sciences at MIT in 1997, and a guest lecturer for the Department of Economics at the Chinese University of Hong Kong in 1998. He also worked on a short-term basis as a visiting professor at Japan's Keio University in 2000, a visiting professor at Harvard University's Kennedy School of Government in 2001, and a visiting fellow at a China research center in France in 2003. These foreign study and work experiences have not only broadened Hu's perspective but also inspired him to promote China studies at home. It has often been said that the field of China studies (*Zhongguo xue*) exists not inside, but outside, China. In the 1990s, as Hu Angang observes, Western scholarship, especially American scholarship, on China studies was far more advanced, in both theoretical paradigms and data-driven empirical research, than its cousin in China.[23]

One of the most important professional objectives for Hu Angang, therefore, is to make China the true center of China studies and of PRC scholarship and to obtain what he calls "the authentic right to speak" (*huayu quan*) on the subject.[24] Hu often tells the Chinese media that PRC scholars cannot engage fully with their Western counterparts until they are well grounded in Chinese studies. Without a doubt Hu is particularly interested in policy issues on the domestic front. Under his leadership, the Center for China Studies has become a leading venue for public policy discourse in the country. The center's policy reports have become one of the best-known sources of policy analysis for the Chinese government. For example, when China was beset with the SARS epidemic in the spring of 2003, the center issued thirty-two reports, which provided policy recommendations on media coverage, public opinion, foreign reactions, the health care budget, and the impacts on the economy and tourism.[25] Hu was also invited to participate in two small roundtable discussions on combating the epidemic, which were held at the State Council and chaired by Premier Wen Jiabao himself.

In 2004, in the aftermath of SARS, Hu Angang wrote a report in which he argues that insecurity in health (*jiankang bu'anquan*) is the largest challenge to China's security and development in the future. According to Hu, China experienced a paradoxical development over the past decade: as per capita income increased, the number of the people with chronic diseases also increased. In addition, Chinese consumption of tobacco and alcohol is the largest in the world. According to an official Chinese study conducted in 2010, China has the world's largest population of smokers (310 million), and every year about one million Chinese people die of smoking-related cancer.[26] In addition, approximately 200 million to 300 million people in China lack access to clean drinking water.[27] Hu's assessment of the health care crisis as China's most daunting challenge has received much attention in the official Chinese media over the past five years.

Hu's foreign studies also broadened his horizons in terms of the role of public intellectuals. He found inspiration from his studies of, and contact with, prominent Western economists. The late Angus Maddison, the prominent British economist, was instrumental in influencing Hu's thinking and research on China's economic development, as evident in this volume. Hu's role model was the late Paul Samuelson, a U.S. economist and a Nobel laureate, who refused to serve as an adviser for the White House. Although Hu enjoys being frequently consulted by Chinese leaders, he has always identified himself as an independent scholar rather than as an aide to a top leader or adviser to the government. "To speak on behalf of poor people," as Hu claims, is the "supreme principle of my professional career."[28] This may also explain why he did not pursue a career working for the Research Office of the State Council or the Policy Research Office of the Central Committee of the Chinese Communist Party (CCP) when such opportunities were presented to him in 1993.[29]

Hu's identity as an independent public intellectual has allowed him to criticize official policies. Hu was one of the first scholars to appeal for the reallocation of resources to China's inland region. In 1994 he made the bold suggestion that China should end special economic zones and abolish preferential treatment for the coastal region.[30] Because of this suggestion, Hu became very popular among officials in inland provinces, while some in the special economic zones accused him of opposing Deng Xiaoping's open-door policy.[31] As some foreign analysts observe, Hu was among the first Chinese scholars to advocate the concept of green GDP growth.[32] According to Hu, China should not measure development merely by "black GDP numbers" but must subtract the immense and variable costs of environmental destruction from impressive GDP numbers in order to measure green GDP growth, a more accurate gauge of real development. While acknowledging the serious demographic challenges that are likely to confront China in the twenty-first century, Hu has steered clear of sensationalist predictions like those of American environmentalist Lester Brown, who posed the controversial question, "Who will feed China?" Hu is reportedly the first scholar to challenge Brown's excessively pessimistic prognosis.[33]

Hu seems able to achieve a balance between maintaining easy access to policymakers and retaining his public image as a credible and independent scholar. The public has sometimes castigated other well-known Chinese economists for their seeming lack of professional integrity for serving only the interests of the rich and powerful, especially vested corporate interest groups. Hu Angang has not earned such a negative reputation. Quite the contrary, some of Hu's most populist policy initiatives were later accepted or implemented by the Chinese government. Examples include:

—Initiatives for an ecologically sound energy efficiency policy (initiated in 1988 and implemented in 2005)

—Policy recommendations (jointly made with Wang Shaoguang) on the tax division between the central and local governments (initiated in 1993 and implemented in 1994)

—An appeal to ban business involvement by the Chinese military (jointly made with Kang Xiaoguang; initiated in 1994 and implemented in 1998)

—An appeal for a more balanced regional development strategy, especially for the need to accelerate development in the country's western region (initiated in 1995 and implemented in 1999)

—A proposal for an employment-centered mode of economic development (initiated in 1998 and implemented in 2002)

—Policy recommendations for reducing or abolishing the agricultural tax on farmers (initiated in 2001 and implemented in 2002)

—Initiatives designed to strengthen China's public health care system (initiated in 2002 and implemented in 2004)

Hu has also long been a strong advocate for political reforms in China. It is important to note that Hu believes that the priority of political reforms should be the development of technocratic decisionmaking in social welfare policy, involving more open discussion and more consultation with think tanks, rather than the adoption of a Western multiparty system, though he does believe in checks and balances on power.[34] In his view, the various provinces in the country—big and small, rich and poor, coastal and inland—should share resources and developmental opportunities. In the mid-1990s Hu proposed, to great controversy, a one-province, one-vote system for the formation of the Financial Committee of the National People's Congress and hinted at a similar method to determine membership in the Politburo of the CCP. In his judgment, this would not only give every province a voice in party policy but also encourage more genuine efforts to ease local dissatisfaction with the central government and ameliorate the disparities between coastal and inland provinces.[35]

Hu Angang has been criticized by his peers in China on various grounds—for example, for his lack of academic disciplinary focus, for a lack of scholarly vigorousness in data selection in the documentation and research methodology in some of his publications, and for inappropriately mixing academic objectivity with subjective nationalistic sentiments. In 2008, for instance, Hu published an 800-page book on Mao Zedong and the Cultural Revolution, which received mixed reviews.[36] In the book Hu offers a more positive view of Mao than is standard, noting his "contributions" to China's rise in the twenty-first century. In Hu's view, China's pre-1978 social and economic development should not be underestimated. Similar views are also expressed in the volume *China in 2020*, in which Hu argues: "From Mao's ideas Deng Xiaoping derived many conceptual innovations concerning reform and opening." Critics will certainly challenge this view by pointing out the sharp contrasts between Mao's revolutionary fanaticism and Deng's economic pragmatism.

Hu's more positive evaluation of Mao and the Mao era, his populist approach that privileges socioeconomic egalitarianism, and his favorable views on the consolidation of state-owned enterprises (SOEs) have made some analysts, both in China and abroad, identify him as a left or new left intellectual.[37] His most biting critics have called him a hack writer of the authoritarian regime, because of his article (co-authored with Hu Lianhe) that criticizes the Western separation of powers.[38] Hu does not apologize for his broad and interdisciplinary approach to China studies. He notes that he is not an academic economist in the conventional sense but instead is one who combines knowledge in economics, politics, culture, and ecology to give a more holistic analysis of a rising China. Hu refuses to be labeled a left, new left, or liberal scholar, arguing that none of these labels accurately reflects his values and worldviews.

As for the accusation that excessive nationalistic sentiment shapes his scholarship, Hu is not defensive. He admits that he is first and foremost a patriotic Chinese citizen, not a seemingly value-free "China hand" (*Zhongguo tong*). As he states in the first chapter of this volume: "This book represents my efforts to observe China as an insider, to understand China as a researcher, to forecast China's future as a participant in its evolution, and as a scholar of the era following Deng Xiaoping to help construct China." In Hu's view, his scholarship should be judged simply by the validity of his arguments and their effectiveness in shaping China's intellectual and policy debates.

Reasons for Chinese Optimism

In general, Chinese citizens are now widely aware that their country is ascendant or, in Hu Angang's words, is en route to the "status of a new superpower." A triumphal mood has begun to take hold in the PRC over the past decade. A series of historic events—China's accession to the World Trade Organization, Beijing's successful hosting of the Olympics, Shanghai's reemergence as a cosmopolitan center as evident in the recent World Expo, the dynamic infrastructure development in both coastal and inland regions, the launch of the country's first manned space program, and the country's ever-growing economic power—have understandably instilled feelings of pride and optimism in the Chinese people.

China's economic strength in today's world can be felt in numerous ways. By the end of 2009 China had foreign reserves of $2.4 trillion, accounting for 30.7 percent of the world's total and making the PRC the largest foreign reserve country for the fourth consecutive year.[39] In the last quarter of 2010 the PRC (including Hong Kong) held 24.3 percent of total U.S. treasury securities.[40] Not surprisingly, as a lead article of *The Economist* observes, "Optimism is on the move."[41] "For the past 400 years," the article notes, "the West has enjoyed a comparative advantage over the rest of the world when it comes to optimism." But now a large number of public opinion surveys consistently reveals a different trend. According to the Pew Research Center study cited in the article, some 87 percent of Chinese think their country is going in the right direction, whereas only 30 percent of Americans do.

Some of the main reasons for Hu Angang's optimistic assessment of China's path toward becoming a superpower—impressive state assets, newly built infrastructure, large domestic market, and advantages in human resources—are empirically well grounded. In this volume, as well, Hu provides a wealth of information and statistics to support his thesis. The four major areas that warrant optimism about China's future are the rise of its flagship, state-owned enterprises; its remarkable development of transportation and infrastructure; its large emerging middle class; and its emphasis on education and innovation.

FLAGSHIP, STATE-OWNED ENTERPRISES

For most of the 1980s and 1990s, and especially on the eve of China's accession to the World Trade Organization in 2001, analysts in both China and abroad viewed the prospects of Chinese state-owned enterprises, particularly Chinese commercial banks, very negatively. This cynical view was understandable because in 1998, for example, bad bank loans accounted for 40 percent of the capital of China's state-owned commercial banks. In 2008, however, the percentage of bad loans among these banks had fallen to 2–3 percent, as Hu Angang notes. The total wholesale profit of China's state-owned enterprises also increased—from 2 percent in 1998 to 7 percent in 2008, more than a threefold increase. Hu was one of the first Chinese economists in the reform era to criticize "market fundamentalism" and favor the consolidation of large SOEs. He argues that the "market is by no means omnipotent, and one should not idealize it."[42]

The recent global financial crisis would seem to support this thesis. Interestingly, in early 2009 four of the world's top ten banks (in terms of market capitalization) were Chinese. The Industrial and Commercial Bank of China, China Construction Bank, and the Bank of China topped the list.[43] A decade ago U.S. banks dominated the top-ten list, and no PRC-based bank was even close to being included. If the economic success of a country is measured by the number of companies that make it onto the Global Fortune 500 list, then the PRC undoubtedly represents one of the greatest triumphs in the contemporary world. The number of PRC companies on the list increased from three in 1995 to forty-six in 2010.[44] The top fifteen Chinese companies on the 2010 Global Fortune 500 list are all SOEs.

TRANSPORTATION AND INFRASTRUCTURE

Over the past decade China has made truly breakneck progress in the transportation and infrastructure sectors, especially in the construction of roads, railways, bridges, and ports. By the end of 2010 the total kilometers of highway in the country had reached 74,000, second only to the United States.[45] China did not have a high-speed railway until 1999, when the Qinhuangdao and Shenyang High-Speed Railways were built, and yet by 2010 the country's total kilometers of high-speed railway had reached 7,400, the highest in the world. It is expected that by 2020, high-speed railways (200 km. per hour) in China will total 18,000 kilometers, which will be more than half of the world total.[46] The world's three longest sea bridges are now all located in China: the Qingdao Gulf Bridge (41.6 km.), the Hangzhou Bay Bridge (36 km.), and the East Sea Grand Bridge (32.5 km.), all completed in the last five years.

As of 2009, of the world's ten-largest container ports, six were located in the PRC (Shanghai, Hong Kong, Shenzhen, Guangzhou, Ningbo, and Qingdao).[47] Two decades earlier, in 1989, none of the PRC's ports was among the top twenty.[48] China's rapid development in infrastructure, especially some of its state-of-the-art transportation projects, are not only the envy of other developing countries but have also become showcases for developed countries such as the United States. This newly built transportation infrastructure will promote commerce, both domestically and internationally, making China more competitive in the world economy in the years and decades to come.

EMERGING MIDDLE CLASS

Hu Angang and other Chinese scholars who are optimistic about China's continuing growth often stress the great significance of the rapid emergence and explosive growth of the Chinese middle class. They credit this expansion with stimulating the country's domestic demand and thus underwriting high growth rates for the foreseeable future. To a certain extent China today is already one of the world's major middle-class markets. In 2009, for example,

China's auto production output and sales volume reached 13.8 million and 13.6 million, respectively, making the PRC the world's leading automobile producer and consumer for the first time.[49]

In his 2010 book, which was based on a large-scale nationwide survey, the distinguished sociologist Lu Xueyi notes that the middle class (based on a definition that combines occupation, income, consumption, and self-identification) constituted 23 percent (243 million) of China's total population, up from 15 percent in 2001.[50] Lu predicts that the Chinese middle class will grow at an annual rate of 1 percent over the next decade or so.[51] Lu also holds that in about twenty years the Chinese middle class will constitute 40 percent of the PRC population, which is on par with Western countries and will make the PRC a true "middle-class nation."[52] According to a study by two analysts at the Brookings Institution, China accounted for only 4 percent of global middle-class spending in 2009 (enough to be the seventh-largest middle-class country in the world) but could become the "largest single middle-class market by 2020, surpassing the United States."[53]

EDUCATION AND INNOVATION

Hu Angang devotes two chapters in this volume to the importance of education and innovation, including the role of science and technology, for China's rise as a superpower. According to Hu, as a large country that lacks natural resources, China should prioritize the development of human resources. Hu and his optimistic colleagues believe that the PRC has, in fact, made tremendous progress in all levels of education. Since 2001, for example, English has been offered in almost all elementary and middle schools in all county-level cities across the country. In the United States, by contrast, the percentage of elementary and middle schools that offer foreign language courses declined from 31 percent and 71 percent, respectively, in 1997 to 25 percent and 58 percent in 2008, mainly due to budgetary constraints. Moreover, only 3 percent of elementary schools and 4 percent of middle schools in the United States offer Chinese courses.[54]

As Hu documents in this volume, the number of Chinese with access to higher education reached nearly 98 million in 2009, 543 times that of the 1949 figure (185,000). In 2009, with 21 million registered college students, "China already surpassed the United States (with 18 million in college) to rank first in the world in terms of students in higher education." In postgraduate studies, for every seven Ph.D. degree recipients in the United States, one is a PRC citizen.[55] Science and technology papers published by Chinese citizens in major international journals in 2008 "accounted for 11.5 percent of the world total, second only to the United States' 26.6 percent." In some research areas, such as life science and biotechnology, nanotechnology, laser technology, and electric cars, China has become a world leader. In 2010 the Chinese leadership issued the *National Medium- and Long-Term Talent Development Plan (2010–2020)*, a blueprint for creating a highly skilled national workforce within the next ten years, including ambitious objectives to recruit and cultivate foreign-educated Chinese nationals (known as returnees).[56]

Challenging Chinese Optimism

Public intellectuals in present-day China are by no means unanimous in sharing the optimism exemplified by Hu Angang's work. In fact, prominent scholars in academic fields ranging from economics to sociology, political science, education, and international relations actively debate both sides of this proposition. This discourse is shaped by their academic writings, policy briefings, media commentaries, blog postings, and participation in semischolarly Chinese websites such as *China Election and Governance* and the Chinese version of *Financial Times* online. These popular websites have apparently survived official censorship despite their bold criticism of Chinese authorities.[57] In general, critics argue bluntly that Chinese optimism is wrongheaded because China's mode of economic growth is neither desirable nor sustainable.

It should be noted that much of this criticism does not directly challenge Hu Angang's arguments but rather serves as a response to the new public discourse on the so-called China model (*Zhongguo moshi*).[58] Nevertheless, these critics tend to believe that optimists' overall prognostications of China's economic and political trajectories (such as Hu's) are highly misleading. They reject the optimists' central argument that China's hybrid system of market economics and authoritarian Leninist statism represents a sound direction for China's continuing rise and even a model for other countries' development. These pessimists' criticism is as broad and comprehensive as the assessment of the proponents of Chinese optimism, covering all of the important economic, social, political, and foreign policy domains.

Critics acknowledge China's impressive economic growth during the reform era, but they believe that it was mainly, if not entirely, achieved through socioeconomic circumstances such as cheap labor, the nature of the early phase of industrialization, technological free riding through imitation, and a lack of concern for energy efficiency or environmental protection. According to the distinguished scholar Cai Fang, director of the Institute of Population and Labor Economics of the Chinese Academy of Social Sciences (CASS), China's "demographic dividend," the term that Hu Angang also uses to refer to a rise in the rate of economic growth due to an increasing share of working-age people in a population, will come to an end in only two to three years.[59] According to Cai, the recent labor shortage in China's southern and eastern cities is a harbinger of far-reaching, important changes for the Chinese economy.

Another critic is the well-known economist Yu Yongding, president of the China Society of World Economics and former director of the Institute of World Economics and Politics at CASS, who has also served as a member of the Monetary Policy Committee of the Peoples' Bank. He argues not only that "China's rapid growth has been achieved at an extremely high cost" but also that the state-led and export-driven growth model "has now almost exhausted its potential."[60] According to Yu, overinvestment in real estate development by local governments and large SOEs—expenditures that now account for nearly a quarter of the country's total investment—is particularly worrisome. As Yu describes, the bursting of the real estate bubble will be not only a shock for "those economists and strategists busily extrapolat[ing] its future growth path to predict when it will catch up to the United States" but also devastating for China, the country that has just "surpassed the threshold for a middle-income country."[61] Yu also argues that some of China's widely perceived competitive advantages are double-edged swords, such as China's large holdings of foreign reserves, which may cause significant instability in the Chinese economy when foreign currencies, such as the U.S. dollar, fluctuate substantially.[62] Yu cautions that it is not impossible that China may eventually fall back to low-income status rather than attain superpower status.

As for the meteoric rise and ever-growing power of China's large SOEs on the world stage, the critics believe that the oligopoly of SOEs not only jeopardizes the commercial interests of foreign companies but also hurts the country's own private enterprises, thus detracting from the long-term potential of China's market economy. This explains the widely repeated new Chinese concept, "The state advances and private companies retreat" (*guojin mintui*), to criticize the growing trend toward "strong government, weak society." A study conducted by Chinese scholars shows that the total profits made by China's 500 largest private companies in 2009 were less than the total revenues of two SOE companies, China Mobile and Sinopec.[63] Ironically, the private sector's net return on investment was 8.18 percent, compared to the 3.05 percent return of SOEs in the country in 2009.[64] The impressive growth of China Mobile has been attributed, at least partially, to the company's monopoly of telecommunications in the Chinese domestic market. With large SOEs monopolizing the telecommunications sector, there is no incentive for these flagship companies to pursue technological innovation. This explains a paradoxical phenomenon: while China's large SOEs have dramatically increased their profitability and standing among the Global Fortune 500 over the past decade or so, no single Chinese brand has truly distinguished itself in the global market. The focus on Chinese innovation seems to be more rhetoric than substance.

Xu Xiaonian, professor of economics and finance at the China Europe International Business School in Shanghai and former managing director of China International Capital Limited, a large state-owned financial company, has emerged as a leading critic of SOE development. Xu uses the Chinese term *quangui zibenzhuyi*, which can be translated as "crony capitalism" or "state capitalism," to express his reservations about the growing trend of state monopoly in present-day China.[65] He believes that with the rapid expansion of SOEs in the past few years, China has in fact begun to reverse Deng Xiaoping's plan for the country's development. In his view, China is drawing the wrong lessons from the recent global financial crisis and heading in the wrong direction.[66] In Xu's view, the main beneficiaries of the SOE growth are corrupt officials, not the Chinese public. Xu believes that in today's China, entrepreneurs only exist in the private sector, not in SOEs, because SOE managers have neither an entrepreneurial spirit nor a sense of responsibility for their companies' losses.[67]

Many critics also believe that business special interest groups have become too powerful. According to Sun Liping, a professor of sociology at Tsinghua University, the real estate interest group accumulated tremendous economic and social capital during the past decade.[68] The power of this corporate interest group explains why it took thirteen years for China to pass an antimonopoly law, why the macroeconomic control policy in the mid-1990s was largely ineffective, and why the widely perceived property bubble in coastal cities has continued to grow. In each of these cases, corporate and industrial interest groups have encroached upon the governmental decisionmaking process, either by creating governmental policy deadlock or by manipulating policies in their own favor.

The Chinese government admits that more than 70 percent of the 121 companies under the State-Owned Assets Supervision and Administration Commission (SASAC), a body overseeing China's largest SOEs, are engaged in the real estate business and property development.[69] These companies also run about 2,500 hotels throughout the country. The official media also criticize these companies for "not doing their proper business" (*buwu zhengye*). According to Xinhua News Agency, the Chinese government ordered seventy-eight SASAC companies to withdraw their investments in the real estate business.[70] Some speculate that some portion of China's stimulus package (4 trillion yuan, or $586 billion) in the wake of the 2008 global financial crisis has been used inappropriately for property development. According to a senior researcher in the Ministry of Housing and Urban-Rural Development, about 32 percent of the stimulus package was invested in real estate.[71] Sun also holds that increasing speculation in property development aggravates the country's economic disparities, making the prospect of joining the middle class even more remote, if not downright impossible, for families of the urban lower class. Sun highlights two parallel and politically dangerous trends that are gaining momentum and may eventually cause a devastating cleavage (*duanlie*) in the country: oligarchy at the top and fragmentation at the bottom.[72]

In contrast to Hu Angang's optimistic view of education and innovation in China, critics point out the various forms of so-called education corruption (*jiaoyu fubai*) that have emerged in the past two decades: the rapid expansion of higher education at the expense of quality, favoritism in admissions, excessive drives for donations and numerous fees, academic inbreeding in appointments and the promotion of faculty, political interference in teaching and research, and plagiarism.[73] Some scholars argue that China's international competitiveness is severely damaged by these problems. One study shows that, among PRC-born doctoral recipients in science and technology in U.S universities in 2002, only 8 percent returned to China by 2007, the lowest rate among all countries (by comparison, India's return rate is 19 percent, Taiwan's is 57 percent, Mexico's is 68 percent, and Thailand's is 92 percent).[74] This study apparently undermines the Chinese government's claim that it has been successful in recruiting foreign-educated Chinese returnees to work in the PRC.

Most important, as many critics argue, innovation-led economic growth can only be achieved in a politically free and open environment, a far cry from the tight political control and media censorship that characterizes present-day China. Liu Junning, a well-known political scientist, is also cynical when it comes to Chinese optimism. He believes that China, when viewed from a historical perspective, is not really on the rise but is "experiencing a fundamental crisis in faith" resulting largely from the prolonged absence of freedom of belief and genuine political reforms in the country.[75]

Some critics believe that the Chinese optimism associated with the China model often leads to an arrogant foreign policy mindset, another departure from Deng Xiaoping's low-profile approach. In a widely circulated article, Zi

Zhongyun, a distinguished international relations scholar and former director of the Institute of American Studies at CASS, launched a bold critique of Chinese optimism, especially the Chinese notion of *shengshi* (magnificent era), a term used by the political establishment to characterize today's China.[76] Zi states bluntly that underneath this superficial "magnificent era" there is a profound sense of crisis in the making and deep concern about the decay of the regime. On the domestic front, as she observes, China's political reform has basically halted, and censorship has almost returned to the horrendous level reached during the Cultural Revolution. On the foreign policy front, what happened in 2010 reminds her of the decade of the 1960s, when China had mostly enemies and only two or three "friends," none of which could feed their people.

This line of criticism of the China model by intellectuals like Cai Fang, Yu Yongding, Xu Xiaonian, Sun Liping, and Zi Zhongyun is notable because none of them are considered to be antigovernment political dissidents. On the contrary, most of them are well-established public intellectuals who may even have close ties with senior leaders in the Chinese government.[77] While some of their views may be a bit sensational, they have forcefully challenged the prevalent feeling of confidence in China today. It is fair to point out that Hu Angang himself does not deserve criticism for some of the problems associated with the China model or Chinese optimism such as economic determinism, crony capitalism, vested corporate interest group politics, or the neglect of demographic challenges. In terms of forecasting demographic challenges confronting China, Hu Angang offered warnings at least as early as any of these critics. Arguably, Hu is also more foresighted and more instrumental than these critics when it comes to pushing for fundamental changes in China's mode of economic growth and protecting the interests of vulnerable social groups. More important, Hu's optimism about China's future is not based on an aggressive foreign policy.

Debating Chinese Exceptionalism

In international affairs, every nation may subscribe to some form of exceptionalism because each has its own unique history, idiosyncratic culture, distinct geography, and specific socioeconomic and political circumstances. An extraordinary characteristic or a combination of characteristics in a given country may lead the country to behave in an exceptional way. In particular, great powers' exceptionalism may have a greater impact on world affairs. For example, American exceptionalism refers to the view that the United States, for a variety of reasons— its absence of feudalism historically, Puritan roots, endowment of a large continent with abundant natural resources, melting-pot ethnic composition, democratic ideology combining republicanism, egalitarianism, individualism, and populism, and its great sense of confidence and responsibility—differs profoundly from other nations, including Great Britain and other European states.[78] While proponents of American exceptionalism differ in important ways, a common argument of this school of thought is that the United States is a lasting "shining city on a hill," which may be exempt from the cycle of power shifts in world affairs and historical forces that have affected most, or all, other great powers.

Chinese exceptionalism (*Zhongguo teshulun*) is a relatively new term used in the Chinese discourse of international affairs, although one may reasonably argue that the Chinese are more aware of their long and often glorious history and their rich and generally distinct culture than any other people in the world. Most of the recent writings on Chinese exceptionalism are related to discussions of the China model, the perceived unique experience of China's economic miracle. In a recent article, Henry Kissinger insightfully observes that the United States and China each "assumes its national values to be both unique and of a kind to which other peoples naturally aspire. Reconciling the two versions of exceptionalism is the deepest challenge of the Sino-American relationship."[79]

In 2004 Kang Xiaoguang, a professor at People's University and former colleague of Hu Angang at the Research Center for Eco-Environmental Sciences of the CAS, wrote a long, scholarly article on this topic and was among the first group of Chinese scholars who used the concept.[80] More recently, Chinese scholars have begun to link the concept to the argument of "China's peaceful rise."[81]

The most influential Chinese proponent of the notion of China's peaceful rise or peaceful development is Zheng Bijian, former vice president of the powerful Central Party School and an aide to many top leaders in the PRC. Zheng argues that China's rise to major-power status will not, and should not, follow the conventional pattern of conflict or war between great powers.[82] China's rise differs profoundly from the conventional pattern of the rise and fall of great powers, he argues, and will not result in a hegemonic conflict, for four major reasons. First, while great power politics in recent world history involved intensified ideological tensions—such as the cold war between the Soviet Union–led communist bloc and the U.S.-led capitalist camp—a rising China does not intend to pursue ideological warfare with the West. In Zheng's words, China is interested in exporting computers, not ideology or revolution.[83] Second, in contrast to Western imperialist powers that established many military bases overseas during their expansion, China will not seek a similarly belligerent and military-oriented foreign policy. Zheng believes that the notion of the peaceful rise could help prevent China from taking the road to militarism. Third, China must learn lessons from Western industrialization, which consumed vast amounts of unsustainable resources and thus will likely pursue an industrial development strategy that is more energy efficient. Fourth, China will not engage in large-scale emigration, which may cause anxieties and problems for other countries, especially China's neighbors.

Hu Angang largely shares Zheng Bijian's strategic vision for China's peaceful rise. In this volume he goes a step further by more systemically examining the domestic impetus and constraints for China's benevolent foreign policy objectives. Based on Hu's analysis, Chinese exceptionalism is attributed to a number of factors specific to China. First, having endured a century of foreign invasions and humiliations, China is guided by the Chinese traditional notion of "not doing to others what you do not want others to do to you." Hu believes that China will neither pursue a hegemonic foreign policy nor impose its own will on others. Second, rather than adopting the industrial development model of the West, with high environmental costs and high energy consumption, China must strive to be a "resource-efficient, environmentally friendly superpower focused on green development." Third, the Chinese leadership's emphasis on domestic development will likely continue as the country prioritizes policies that address economic disparities, the lack of public health care, energy inefficiency, and ecological challenges. This domestic and people-centric approach will prevent the country from pursuing a belligerent foreign policy. Fourth, the development of human resources has been the most important impetus of China's rise. For the same reason, China's superpower status ultimately depends on a "transition from a knowledge-closed nation into a knowledge-open nation and from a country lagging in science and technology into a country spearheading global innovation." The age of a knowledge economy, as Hu characterizes it, should be "defined less by competition than by cooperation." Hu concludes, therefore, that "China's rise represents an opportunity for the rest of the world rather than a threat."

Foreign analysts, of course, are not so naïve as to confuse promises with reality. The true objectives of Chinese foreign policy must ultimately be judged by actions rather than the words of Chinese leaders and their advisers like Zheng Bijian and Hu Angang. Some Western critics may even argue that these notions of peaceful rise and Chinese exceptionalism are nothing but diplomatic gestures on the part of Chinese leadership to temper the concerns of PRC's neighboring countries and other major powers about the emergence of a militarily strong and expansionist China.

It is important to note, however, that there is another school of thought criticizing the naïveté associated with the notion of China's peaceful rise that is increasingly gaining momentum in the PRC. Ever since Zheng articulated his concept Chinese critics have been cynical about its acceptance both domestically and abroad.[84] For example, Yan Xuetong, a well-known political scientist at Tsinghua University, observes that the notion of China's peaceful rise is confusing and ineffective due to three contrasting interpretations of the terms *rise* and *peaceful*. The first interpretation considers both *rise* and *peaceful* as goals and, therefore, holds that China seeks an exceptional way to rise, which differs from the many powers in history that rose to prominence through military means. The second interpretation treats *peaceful* as a means and *rise* as an end and, therefore, suggests that China uses *peaceful* in name only in order to achieve its desired end: to rise. The third interpretation regards *peaceful* as an end and *rise* as a means and, therefore, understands China to be seeking to contribute to world peace.[85] According to Yan, tensions among these meanings will immediately emerge in the policymaking process, because no government will be so rigid as to adhere to one promise when international politics is an ever-changing game. Thus, as Yan implies, the concept of a peaceful rise is no more than wishful thinking. One military critic, Major General Luo Yuan, criticizes pacifism (*heping zhuyi*) and the pan-pacifist approach (*fan hepinghua*), which are popular among some members of the Chinese foreign policy establishment.[86] Luo explicitly argues that this pacifist thinking will lead China to follow the path of post–World War II Japan, falling into a trap, set by the United States, that prevents it from becoming a superpower.

The strongest and most explicit critique of Chinese exceptionalism is the 2010 bestseller, *The China Dream*, written by Senior Colonel Liu Mingfu, director of the Institute of Military Development of the National Defense University. Liu explicitly argues that China should pursue a new developmental strategy of "military rise" (*junshi jueqi*) to obtain and secure a global leadership position from which it can compete with the United States.[87] Liu believes that China's rise fits within the pattern of shifts in power in the international system, and he therefore rejects the idea of Chinese exceptionalism. In his analysis, a hegemonic conflict is inevitable. Liu Mingfu's book is not based on serious scholarly research but rather on snapshots of historical events and random references to the views expressed by some Chinese and foreign strategic thinkers. A majority of Chinese leaders and public intellectuals may not agree with Liu's arguments and strategic perspective, but Liu's view does represent a subset of sentiments among some international relations scholars, opinion leaders, and especially military strategists.

In the book Liu Mingfu explicitly states that the overall goal of China's modernization is to become "the most powerful country in the world" (*touhao qiangguo*).[88] To justify this strategic goal, Liu argues that all of China's paramount leaders over the last century shared this goal. In the 1911 Revolution, Sun Yat-sen said that his dream was to build China into "the number-one richest and strongest country in the world" (*shijie diyi fuqiang zhiguo*); Mao's Great Leap Forward was nothing but an attempt to surpass the United States; and Deng Xiaoping's "keeping a low profile" strategy (*taoguang yanghui*) strived to achieve, in three steps, Chinese national rejuvenation above all others by the middle of the twenty-first century.[89]

According to Liu, China's replacement of the United States as the sole superpower, or what Liu labels the "superpower succession" (*guanjun guojia gengti*), is not far in the future. He asks Chinese leaders and the public to remember that China should not have any "illusions" about American goodwill regarding "China's rise" or about peaceful coexistence between the existing superpower and the emerging superpower. In his view the United States is determined to contain China economically, politically, ideologically, and, most important, militarily.

Liu's analysis is widely shared by other analysts in the Chinese military establishment. Minister of Defense Liang Guanglie's remark that "peace does not fall from the sky" echoes Liu's argument for the need to accelerate China's military modernization.[90] The ultimate means for China's rise, especially according to military analysts, is to build strong armed forces to compete with the United States on that front. Liu argues that this is not as much because China wants to defeat the United States as because the United States wants to defeat China.[91] Becoming a military superpower, as Liu argues, should be the "Chinese dream," without which China will have only "nightmares."[92]

These debates over Chinese optimism and Chinese exceptionalism illustrate the importance of Hu Angang's work—its originality, complexity, foresightedness, and comprehensiveness. They also reveal the diversity within China's intellectual community. Each reader, of course, can arrive at his or her own judgment of the author's analyses, the prospects of a rising China, and its implications for peace and prosperity in the twenty-first century. It is reasonable to assume, however, that the ideas, debates, and visions competing in China matter not only for China's own future development but also for its interaction with the outside world. As a champion for Chinese optimism and exceptionalism, Hu Angang can increase our knowledge and enhance our understanding of the pressing issues and long-term challenges that face this rapidly changing country. Considered from an even broader perspective, this intriguing and forward-looking book may also contribute to international recognition that in the new global environment, we all must find novel ways to think about power and responsibility.

CHAPTER ONE

CHINA, AN EMERGING SUPERPOWER

O ne of the great events of the past three decades has been the rapid rise of the People's Republic of China (PRC). China's quick ascent into the ranks of great powers not only outstripped the expectations of the international community but has also far surpassed the Chinese government's own expectations.[1] If calculated using official constant prices, China's aggregate GDP in 2009 was 18.6 times that of 1978, meaning that the economy averaged an annual growth rate of 9.9 percent over that thirty-one-year time span. Over the same period, per capita GDP grew at an average of 8.7 percent a year, leaving it 13.3 times larger in 2009 than it was in 1978.[2] If we employ the exchange rate method of calculating GDP, China is the second-largest economic power in the world, falling behind only the United States.[3] Purchasing power parity (PPP) calculations of GDP place China second only to the United States.[4] If economic clout is assessed according to merchandise exports, then China surpassed Germany in 2009, to assume the number one position.[5]

Even though China's import and export throughput has decreased and its economic growth in general slowed in the wake of the global financial crisis, there has been no change to the country's general trend of industrialization and urbanization. Indeed, the forces of neoliberal market economics, in tandem with the information revolution of modern globalization, have continued to sweep across China, paving the way for a strong 9.1 percent economic growth in 2009. With much of the world still mired in economic decline, China was the only country to attain such a high rate of GDP expansion.

Clearly, China's rate and scope of development merit the superlative "economic miracle" so often bestowed upon it by outside observers.[6]

Despite China's successful development over the past thirty years, skeptics never tire of questioning the country's economic health. Such fallacies as "China's GDP bubble" and "the China collapse theory" have been popularized by those who doubt the foundations and the sustainability of the PRC's economic development. The tremendous success of China's political, economic, and social development has also aroused great concern in the West, where one scholar (reflecting the opinions of many) regards China as "the biggest potential ideological competitor to liberal democratic capitalism since the collapse of Soviet Union."[7] One Western analyst goes so far as to claim that "what China has achieved in the last couple of decades legitimately lays siege to many of our most deeply held notions about the realities of government and economics."[8]

Certainly China's transformation has generated a host of questions: Is China really rising? Can China attain a sustainable position as an economic powerhouse for the years and decades ahead? How might this be accomplished? What is the difference between the economy of China and developed economies such as those in Europe, the United States, and Japan? Will China soon emerge as the world's newest superpower? If so, how will this change the world's economic and political landscapes? This book attempts to answer these and related questions.

This opening chapter briefly surveys a variety of forecasts of China's future formulated by institutions and scholars outside of the PRC. It also explicates how to interpret the trajectory of China's economic development and provides an explanation for China's rapid rise. Finally, it provides a brief overview of the book's remaining chapters.

Forecasting China's Future: Foreign Perspectives and Literature Review

Over the past three decades China has experienced the most rapid economic rise the world has seen. In 1983 the World Bank issued its first report on socialist China.[9] The report predicted that China's economy would maintain an annual growth rate of about 4–5 percent in the first half of the 1980s, expanding at a clip of 5–6 percent (the world average in the 1970s) during the latter half of the decade. Two years later the World Bank released a second report, titled *China: Long-Term Development Issues and Options*.[10] This time, the World Bank predicted that China's economy would grow at a rate of 5.4

to 6.6 percent between 1981 and 2000. In reality, China's economy grew at an annual rate of 9.9 percent during that period, far exceeding these estimates. As of the year 2000 China's total GDP was 78.5 percent larger than predicted by the World Bank.[11] In other words, the World Bank had vastly underestimated China's potential for economic growth.

In 1986 John King Fairbank, a professor at Harvard and a leading American sinologist, posited that from a historical perspective China was likely to continue its pattern of following the successful establishment of a new governing authority with a period of great construction.[12] His words proved prescient. After the first generation of leadership, led by Mao Zedong, founded the PRC, the next generation of leadership, led by Deng Xiaoping, launched an ambitious program of economic reform and opening. Deng's efforts ushered in a still-unfurling period of great construction and development.

In 1987 the Yale professor and famous American historian Paul Kennedy expressed great optimism regarding the future of China's development. In his book *The Rise and Fall of the Great Powers,* he noted that China was both the poorest of the major powers and a country not particularly well situated strategically. And yet, despite these challenges, he recognized that Chinese leaders were in the midst of implementing a grand, ideologically consistent, and far-sighted strategy. As a result, he predicted that if China could maintain 8 percent GDP growth, tremendous changes could take place in the ensuing years.[13]

In 1988 President Richard Nixon, in his book *1999: Victory without War,* predicted that the world would be defined by a multipolar political, economic, and military structure after the collapse of the cold war system of bipolarity.[14] He declared that after the onset of the twenty-first century the dominant positions of the United States and the Soviet Union would decline, allowing the three geopolitical giants Europe, Japan, and China to rise. He also noted that the reforms initiated by Deng Xiaoping had liberated the huge potential of the Chinese people, a group that constituted a fifth of the world's population. He then boldly predicted that if China continued along the road paved by Deng Xiaoping, the world would eventually enter an era with three superpowers: the United States, the Soviet Union, and China. Later developments would prove Nixon right in some respects but wrong in others. The importance of geopolitics for Europe and Japan actually decreased, and the Soviet Union dissolved. Indeed, by 2008 Russia's economic strength was only 14.4 percent of China's and 13.5 percent of that of the Unites States.[15] With the decline of the Soviet Union, the United States emerged as the world's sole superpower. In the long run, however, Nixon's predictions relating to China proved prophetic, as it rose to emerging superpower status in the early years of the twenty-first century.[16]

Through the 1990s China did not halt reform and opening because of the June 4 incident at Tiananmen Square. Nor did the drastic changes in Eastern Europe and the disintegration of the former Soviet Union prompt China to change its political orientation. Instead, China continued to build economic momentum. Over time, this aroused concern in many quarters and inspired members of the international community to make a variety of predictions regarding China's future. For instance, a Rand Corporation report released in 1995 predicted that China's GDP would grow 4.9 percent annually from 1994 to 2015, surpassing the United States in 2006 and growing to 1.27 times that of the United States by 2015 (using PPP calculations as the standard). Per capita GDP would also grow but would only reach 28.9 percent of U.S. per capita GDP.[17]

In 1997 the World Bank issued another report, *China 2020,* which consists of a very bullish appraisal of China.[18]

China is in the midst of two historic transitions: from a rural, agricultural society to an urban, industrial one and from a command economy to a market-based one. The interplay and synergy between these two transitions have sparked rapid growth. The Chinese economy expanded more than fourfold in the past fifteen years. In the rich industrial economies this transition took centuries. In China the process is being telescoped into one or two generations.[19]

The report calls China's transformation "the most remarkable development of our time." In this volume the World Bank admits that the projections of its past reports tend to underestimate growth, often substantially. The report goes on to predict that Chinese GDP growth in the period 1995–2000 would be 8.4 percent, with growth from 2001 to 2010 averaging 6.9 percent a year.[20]

Yet again the World Bank underestimates China's growth potential. In the years 1995 to 2000 China's GDP grew at an average annual rate of 8.5 percent, increasing to 10.7 percent between 2001 and 2009.[21] If GDP is computed using the PPP method, China's total GDP will far outpace the United States by 2020, with per capita GDP reaching half the U.S. figure by then. In addition, China's total purchasing power could overtake Europe's and may begin to compete with industrialized countries as a capital provider and recipient in the world's financial markets.[22] In 2009 China surpassed Germany to become the world's largest exporter. It also became the world's second-largest importer, trailing only the United States.

In 1997 the Asian Development Bank published *Emerging Asia: Changes and Challenges,* a report outlining three possible scenarios for China's future economic development.[23] According to the optimistic scenario, China's per capita GDP would grow by 6.6 percent annually from 1995 to 2025. The pessimistic scenario held that per capita GDP would grow at 4.4 percent over this period. The middle-of-the-road scenario predicted that per capita GDP would grow at 6.05 percent annually. The report concluded, based on the middle-of-the-road scenario, that by 2025 China's per capita GDP would be 38.2 percent of U.S. per capita GDP.

Not all foreign growth projections for China have been wrong. In 1998 the British economic historian Angus Maddison made what he called "a considerably conservative prediction" regarding China's future growth, stating that China's annual GDP (PPP) growth would drop from 7.5 percent, the 1978–95 average, to 5.5 percent for the period 1995–2010.[24] He also predicted that per capita GDP growth rates would drop to 4.5 percent a year, from 6.04 percent. Despite this slowdown, he projected that China's aggregate GDP would surpass that of the United States by 2015 and come to account for 17.4 percent of the world total (with U.S. GDP accounting for 17.3 percent). Few Chinese academics found his claims credible. To my knowledge, Maddison is the only scholar who accurately forecast the year that China's total GDP would overtake that of the United States.

In 2001 Goldman Sachs's chief economist Jim O'Neill coined the term BRIC, an acronym referring to the emerging powers Brazil, Russia, India, and China.[25] In a 2003 publication O'Neill predicted that China will be the world's largest economy in 2045, followed by the United States and India.[26] The 2001 paper calls China a "global factory," describing it as an economic entity with all of the necessary factors for strong economic development. First, it imports the largest amount of foreign capital; second, it has become a production base for the largest enterprise groups in the world; third, it has a huge, cheap, and reliable labor force; fourth, the human capital of this labor force improves over time.

Nevertheless, not all of Goldman Sachs's predictions are rosy. In two books following the BRIC papers, Goldman predicts that China's GDP growth will drop to 5 percent annually by 2020 and by 2040 might hover around 3.5 percent.[27] Despite this decline in growth, the nation's high investment rate, huge labor force, and steady integration into the global economy will still ensure its becoming the world's largest economy by 2041. The other BRIC countries are also expected to climb the ladder of economic prosperity. A Goldman report in December 2009 predicted that the BRIC group will overtake the G7—the

United States, Japan, Germany, Britain, France, Italy, and Canada—in terms of economic power.[28]

In 2004 the United States government's National Intelligence Council (NIC) issued a very influential report, *Mapping the Global Future.*[29] One of the main components of the report is an evaluation of China's and India's potential to become important new global players by 2020. The document concludes that the likely emergence of China and India, as well as others, as major global players would transform the geopolitical landscape on the scale of Germany's nineteenth-century unification and America's twentieth-century ascendance to superpower status. A significant difference, however, is that the former developments took place exclusively in the West, while the latter would be distinctly non-Western. In completing the report the NIC surveyed more than a thousand American and foreign experts with a variety of backgrounds: academics, businesspeople, government officials, and members of nongovernmental organizations and other institutions. Most forecasts hold that the dollar value of China's GNP will overtake that of any single Western economic power except the United States by 2020, while the value of India's output could match that of a large European country. Because of the sheer size of China's and India's populations, their standards of living need not approach Western levels for them to become important economic powers. In other words, despite having per capita income levels that may be only a fraction of those in the developed world, China and India will become world powers by virtue of their size and economic clout.

In 2006 the Center for Strategic and International Studies and the Peterson Institute for International Economics released *China: The Balance Sheet.*

> Many aspects of China's economic picture are impressive, even amazing. Already, China is the world's fourth-largest economy and third-largest trading nation. It has grown by about 10 percent per year for almost three decades, increasing its output by a factor of nine since launching its economic reforms in 1978. In the process, it has lifted more than 200 million people out of poverty. . . . In recent years it has accounted for about 12 percent of total growth in world trade—much more than the United States.[30]

The book's authors also observe that China is the second-largest recipient of foreign direct investment and uses its "deep and rapid integration into the world economy" to overcome internal resistance to continued economic reforms.

In July 2008 the Carnegie Endowment for International Peace's Albert Keidel predicted that China would match the United States in terms of economic size in 2035 and that by 2050 China's GDP ($82 trillion) would be nearly two times that of the United States ($44 trillion).[31]

In 2009 Angus Maddison made yet another forecast, projecting China's per capita GDP (PPP) to grow 4 percent annually from 2006 to 2030. Maddison reaffirms his earlier prediction that the PRC's aggregate GDP will overtake that of the United States by 2015. By 2030 he expects China's economy to be one and a half times larger than that of the United States.[32]

To a great extent the rise of China took most members of the international community by surprise. According to Gregory C. Chow, a professor of economics at Princeton, "The country today is so different from what it was in 1978 that hardly anyone in or outside China in the late 1970s could have expected such dramatic changes."[33] Many international institutions and scholars found it very difficult to accurately forecast not only China's actual economic growth but also the forces driving it. Thus far China's rapid economic growth has lasted thirty-two years. The economic rise of the United States, Japan, and South Korea lasted forty-three years (1870–1913), twenty-three years (1950–73), and thirty-one years (1965–96), respectively. How long will China continue to grow? What will China rely on to drive and maintain its growth? These questions remain unresolved. In 1997 the World Bank questioned whether China could maintain its rapid growth, stating that the strength of the Chinese economy over the past two decades did not guarantee that China would continue to grow rapidly in the future.[34] The study suggests that if certain problems could not be resolved they would be likely to harm future growth and cast a shadow over China's future.

How is it that China has maintained over the preceding thirty years such high levels of economic growth? One major reason is that it has avoided the economic fluctuations and incidents of protracted political turmoil, such as the Great Leap Forward and the Cultural Revolution, experienced in the late fifties, late sixties, and early-to-mid seventies. In addition, the PRC has been able to avoid the sort of national disintegration and social conflicts inspired by Soviet-style reform, maintaining social order and stability while concurrently carrying forward the basic principles of Chinese-style reform and opening. A stable regional and international environment has also aided China in its transformation. Finally, China's economic, political, and national defense capabilities have been significantly enhanced through the process of reform and opening, leading the country to become increasingly important in international political and economic affairs.

The Rise of Modern China: This Author's Perspective

Inspired by the three-step strategy advanced by Deng Xiaoping, I divided China's contemporary economic development into three stages in my 1989 book *Population and Development*, with the first beginning in the 1950s and the third ending in the middle of the twenty-first century.[35]

The first stage, from 1950 to 1980, is what I term the preparation-for-growth stage. In this stage, China's social and economic conditions were not yet ripe for industrialization and modernization, and per capita income levels, as well as per capita GNP levels, remained very low. The preparation-for-growth stage is basically the period of time during which the necessary preparations are made for the launch of the contemporary phase of economic development.

The second stage in China's economic development, lasting from 1980 to 2020, is what I call the high-speed-growth stage. In this stage, per capita income rises from the low-income level to the middle-income level, with per capita GNP rising from $300 to $1,700–$1,800 (1980 price). As industrialization gains momentum, the economic aggregate is rapidly augmented. The overall structure of the Chinese economy also changes markedly, with half of the total transition completed. Over time the proportion of the total workforce employed in the nonagricultural sector comes to exceed 50 percent, along with a growth in the proportion of the population residing in urban areas. Profound changes occur in the country's social development, and a new pattern of opening up to the outside world takes shape, integrating China ever more closely with the global economy. In this stage China steps into a new era marked by historically significant economic and social transitions. This high-speed-growth stage allows China's economic takeoff and is the crucial backdrop for the overall revival of the Chinese nation.

In *Population and Development* I projected China's growth from 1985 to 2000 based on six postulations:

—The reform and opening principles that guided Chinese development since the third plenum of the Chinese Communist Party's Eleventh Central Committee would continue to serve as the preeminent economic doctrine.

—There would be no great strategic policy blunders, such as the Great Leap Forward, that would cause significant economic fluctuations.

—The political turmoil and social disorder that characterized the Cultural Revolution would not be repeated.

—The problem of rapid population expansion could be resolved.

—There would be no natural disasters serious enough to affect the national situation.

—There would be no large-scale military confrontations between China and other countries.

These postulations are interconnected and interdependent, with a change to one able to impact the others.

At the time, I estimated that the Chinese economy would grow at an average of 5.9 percent annually from 1985 to 2000—and continue to grow at an average of 4.8 percent until 2020. The medium- and long-term development goals I outlined called for the enhancement of China's economic, political, and national defense power to levels such that by 2020 China could serve as one pole in a multipolar system of global power that would include the United States and the Soviet Union; in this system, China would have the ability to operate independently and to have a decisive say in Asian affairs.

The third stage of development, lasting from 2020 to 2050, is what I call the steady-growth stage. If during this period China were able to attain sustainable, high-speed economic growth, then per capita GNP could range from $1,700 to $13,800, with overall GNP growing at a rate of 4–5 percent. China's GDP would then settle into the world's number-one position, and its per capita GNP could exceed middle-income countries' average. This is my basic prediction of China's development stages: it would have China follow an economic trajectory quite different from other powers.

In 1995 I used the Belgian economist Paul Bairoch's data on various countries' manufacturing totals as a proportion of the world aggregate to explain the rise and fall of major powers from 1750 to 1980.[36] Throughout history, the rise and fall of great powers has been a subject of great concern. Each major power has experienced its own unique trajectory of development, with some moving from weak to strong and others moving from strong to weak. For example, the United Kingdom's growth curve is that of an inverted U. In 1750 its manufacturing accounted for a meager 1.9 percent of the world's total. By 1880, however, that number had risen to 23 percent, making the United Kingdom the world's most powerful manufacturing center. About a hundred years later the United Kingdom had relinquished its place at the top of the manufacturing ladder, accounting for just 4 percent of worldwide production.

Unlike the United Kingdom, China's development curve takes the form of an upright U. According to calculations by Paul Bairoch, by 1750 China's manufacturing made up a third of the world's total.[37] Rapid declines in national strength between 1800 and 1950, however, led to a corresponding drop in manufacturing output. In 1900, 6.0 percent of worldwide manufacturing was accomplished in China. By 1950 only 2.3 percent of manufacturing was done in China. This precipitous drop in production can be explained by the historic circumstances China faced: between 1800 and 1950 China was the victim of several foreign attacks and watched as its territorial holdings shrank. In 1995 I estimated that China would shift from weak to strong between 1950 and 2050, completing the U curve. Thus far, my projections have proven accurate.

In assessing the economic power of China, I used three metrics: official exchange rate, PPP, and physical economy. When I speak of physical economy, I refer mainly to China's agricultural output as a percentage of the world total. In 1995 China had already overtaken the United States to become the world leader in the production of cereals, cotton, peanuts, rape seeds, tobacco, meats, chicken, eggs, and aquatic products. In terms of overall manufacturing, I expected China to catch up with the United States and Japan in the 1990s.[38] But during the 1990s China's annual output of motor vehicles was only 1 million, far smaller than that of the United States and Japan. But in terms of the economic aggregate's physical manufacturing, it could overtake the United States.

My study shows that China was the beneficiary of the type of timing, geographic location, and popular sentiment necessary for an economic take-off and eventual ascendancy to superpower status. History is now providing China with the same opportunity to thrive that the United States enjoyed between 1870 and 1913, Japan between 1950 and 1973, and Korea between 1965 and 1996. External conditions—such as East Asian peace and stability, the information revolution of the 1990s, and China's accession to the WTO—provide a platform for China to take off economically. Domestically, China enjoys popular support for its development. By this I mean that China has been able to maintain social and political stability for an extended period of time and that the thesis "Development is a hard truth," famously explicated by Deng Xiaoping, has become the social consensus.[39] The popular desire for stability has been a vital force in maintaining the nation's cohesiveness and inspiring its economic takeoff.

I also analyzed the basic factors favorable to China's economic takeoff from a purely economic perspective. First, I looked at stock capital and sources of economic development. China maintained fairly high rates of investment savings and foreign investment, beginning with reform and opening, and these continue to stimulate China's economic development. Second, I looked at

China's industrial foundation, concluding that the country had already built a fairly complete industrial system. Third, I examined social infrastructure. Here, I noted that China had made considerable progress in the realms of public transportation, postal service, telecommunications, and urban utilities. Fourth, I looked at the domestic market, observing that China's huge market represented a significant advantage. Fifth, I noted that China's advantages in terms of human resources were pronounced, with its abundance of cheap labor and a relatively well-educated population. All of these factors conspired to make China's 1980s economic takeoff different from that of other countries. This led me to predict that if the modernization and economic development process were not interrupted, China would become a major economic power at the beginning of the next century. I went on to project that China would become a superpower by 2020—politically, militarily, and in terms of science and technology.[40] As of the mid-1990s, this was my perspective on China's development. In the years that followed I continued to pursue this general line of reasoning, updating the data and elaborating my views.

In 1999 I began to employ Angus Maddison's compilation of global economic data and major research findings in my studies of China's development trajectory.[41] Using Maddison's data set, I was again able to show that the historical trajectory of Chinese development takes the form of a U curve, from strong to weak and then from weak to strong.[42] According to Maddison's calculations, China's GDP was the largest in the world in 1820, constituting 32.9 percent of the global aggregate. By 1950, however, the newly established PRC's GDP comprised only 4.5 percent of the world total. In other words, over the course of the 130 years between 1820 and 1950, the size of the Chinese economy dropped from one-third of the world total to less than one-twentieth. This is the downward slope of the U curve. From 1950 to 1978 China's share of world GDP remained below 5 percent. After 1978, however, GDP growth rates began to ascend rapidly. This period, which continues through the present day, drives China ever closer to completing the U curve. As China's GDP as a percentage of the world total began to climb the U curve, Western Europe, the United States, and Japan began to traverse the downward slope of their inverted U-curve trajectory of development (figure 1-1).[43]

With this as a backdrop, this book discusses how China rose so rapidly to become an emerging superpower. My main thesis is that, from the perspective of the history of modern global economic development, China is due for a period of rapid GDP growth. Moreover, China's rise represents an opportunity for the rest of the world rather than a threat. Simply put, China's rise is

Figure 1-1. *Gross Domestic Product, Four Major Powers, Selected Years, 1–2030*

Percent of world total

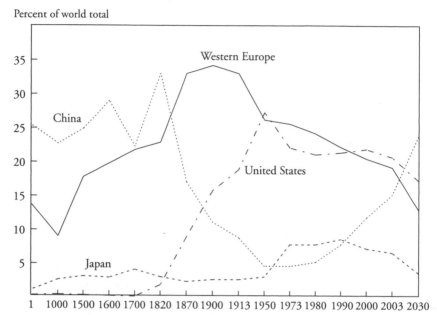

Source: Angus Maddison, "Statistics on World Population, GDP, and per Capita GDP, 1–2003 AD" (www.ggdc.net/maddison/); Angus Maddison, *Chinese Economic Performance in the Long Run, 960–2030 AD* (Paris: OECD, 2007).

not only transforming its domestic political, economic, and social landscapes but also reshaping the world.[44]

China as a New Type of Superpower

When will China become a superpower? What kind of superpower will China be? The first question is easy to answer. Indeed, this book provides a concrete projection for China's ascendancy to superpowerdom based on the country's development. The second question is much more difficult to answer. Different people have different views. This book predicts that China will be a mature, responsible, and attractive superpower.

Before discussing when China will be a superpower and what kind of superpower it will be, one must first delineate the meaning of the term *superpower*. There is no clear consensus on the definition of the term, but people tend to base their judgment on the following factors:

—Culturally, the superpower must be tolerant, making possible the survival and development of all civilizations.

—Geographically, it must feature vast land or sea territories.

—Economically and financially, it must be extraordinarily powerful.

—Demographically, it must have a large number of well-educated citizens and a well-developed infrastructure.

—Militarily, it must possess a unified military power that is relatively indestructible, capable of preventing or causing huge destruction, and able to project its influence across the globe.

—Politically and ideologically, it must have a powerful political system capable of efficiently allocating resources to realize its global political objectives and to exert influence via ideology.

Of course, these six factors do not exhaust the possible factors that make a country a superpower. In my opinion it is more important to be able to identify which countries are superpowers than it is to provide a detailed and comprehensive definition of the term. Without a doubt, the only superpower in the present-day world is the United States. If China overtakes the United States in several important ways, then China will also be a superpower. With this in mind, I am less interested in specifically defining the term *superpower* than I am in analyzing whether or not China can surpass the United States in certain important realms.

For China to become a superpower it must catch up to and surpass the United States in a variety of ways, and this will probably take dozens of years. The key metrics I am referring to are economic power, human resources and human capital, science and technology achievements, and the ability to address climate change. That China is trending in this direction is readily apparent. Over the past thirty years China has succeeded not only in dramatically reducing the numbers of what was once the world's largest population of individuals living in poverty but also in transforming itself into a global economic power. Between now and 2020 China will continue to increase its economic strength until it attains superpower status, thereby ending the era of American hegemony. The international balance of power, then, will shift from unipolarity to a multipolar world dominated by the United States, China, and the European Union. In other words, the world seems, yet again, poised to enter an era of great powers.

Unlike the previous era of great powers, however, the coming age will be defined less by competition than by cooperation.[45] Due to the steady progress of economic globalization, the world's powers are increasingly economically integrated, with their shared interests far outweighing potential areas of conflict. Even without China's arrival, the world would not be dominated by a bipolar, Soviet-U.S., system of checks and balances characteristic of the cold war. Joint rule of the world by the great powers will become the system of the day. Suggestions of this trend can already be found in the Group of Eight (G8) summits held in Washington in November 2008, London in April 2009, and Toronto in June 2010. There, leaders of the world's major powers got together to discuss major political and economic issues as well as climate change and environment protection. As an emerging superpower, China is increasingly integrating into the international community. Just as China's problems have the potential to affect the international community, problems within the international community have the potential to affect China. Such problems require international cooperation in order to reach a resolution. Thus it is necessary for China to institutionalize strategic dialogues with other powers, such as the United States, Japan, the Shanghai Cooperation Organization, and the European Union so as to enhance mutual trust and increase their shared capacity to cope with problems and challenges of an international nature.

Mao Zedong was the Chinese leader who first came up with the strategic concept of catching up with and then surpassing the United States. In August 1956 he told the first session of the preparatory meeting for the Eighth National Party Congress that China must catch up with the most powerful capitalist country, the United States. His rationale was the following:

> The United States has a population of only 170 million, and as we have a population several times larger, are similarly rich in resources, and are favored with more or less the same kind of climate, it is possible for us to catch up with the United States. Oughtn't we to catch up? Definitely yes. The United States has a history of only one hundred and eighty years, and sixty years ago it produced 4 million tons of steel, so we are sixty years behind. Given fifty (2006) or sixty (2016) years, we certainly ought to overtake the United States. This is an obligation. You have such a big population, such a vast territory, and such rich resources, and what is more, you are said to be building socialism, which is supposed to be superior; if after working at it for fifty or sixty years you are still unable to overtake the United States, what a sorry figure you will cut! You should be read off the face of the earth. Therefore, to overtake the United States is not only possible but absolutely necessary and obligatory. If we don't, we the Chinese nation will be letting the nations of the world down, and we will not be making much of a contribution to mankind.[46]

It should be noted that Mao Zedong and Deng Xiaoping both dreamed of making China a strong power but never stressed that China would be a superpower.[47] They did not mean to suggest that China should not be a superpower but that China should never be a hegemonic power that bullies smaller countries. Ultimately, it was the common desire of the two leaders to better China's contributions to the world. At this point it is not necessary to discuss whether or not China should become a superpower. What should be focused on is what kind of superpower China should become.

China should not be a superpower that seeks hegemony and world domination. It should be a superpower that advocates for equality in international relations, cooperation, and collaborative rule of the world. It should be a superpower that works to benefit all parties and interests instead of seeking to rule supreme over all others. It should be a superpower that advocates multilateralism and tolerance instead of unilateralism and exclusion. With more than a century's history of being invaded, bullied, and humiliated by other powers, China knows clearly the meaning of unjust and unequal international relations. As a country that is guided by the Chinese traditional notion of not doing to others what you do not want others to do to you, China will never be belligerent to other nations and will never "deal with others as they deal with you." On the contrary, China is more inclined to resolve contradictions, make concessions, accommodate others, and engage in mutual exchanges. China favors independence. It will never interfere in the internal affairs of other sovereign nations. In terms of domestic development, China promotes modern management featuring transparency and good governance, which is different from the democratic management of the United States or traditionally centralized autocratic systems. China is not inclined to allow leaders directly elected by the people to govern the country. Instead, it stresses public participation and increased governmental transparency.

China's rise has been defined by peaceful development. Throughout this process, China and the international community have been mutually dependent. For this reason, China must consider both the interests of the world and its own interests. Thus that which conforms to the interests of China today tends to be conducive to promoting the interests of the world as a whole. This is why China advocates for a unity of nationalism and globalism rather than putting nationalism above globalism. China does not want to replace the United States and become the sole leader of the world. Rather, China needs to cooperate with the United States to cope with global challenges in economics, politics, energy, and the environment.

China should be a resource-efficient, environmentally friendly superpower focused on green development. The world's industrialized countries, in their modern development, share three features: high resource consumption, especially of nonrenewable resources; high rates of pollution; and high per capita consumption of goods. These are the basic features of the traditional model of modernization as experienced by countries such as the United States. The prerequisite for such traditional development patterns is the availability of certain resources. During their development, these high-income countries, with less than 20 percent of the world's population, were responsible for 75 percent of resource consumption. China does not have such a luxury. Given the size of its population, the resources available to it, and its overall level of development, China must employ a nontraditional modernization model, especially as regards resource use.[48] In doing so, it should establish a resource-efficient and environmentally friendly national economic system featuring low per capita energy consumption rates and minimal pollution. Such actions will help to conserve resources and raise their utility rate, reduce energy consumption, and increase the country's general productivity and population-bearing capacity.

China should be a superpower with a high human development level and a small gap between rich and poor. In recent years, its guiding principle for modernization has shifted from being economy centered to people centered, with human development gradually becoming the first and foremost objective of development. At this point most of the Chinese population has attained a high level of human development, and the standard of living continues to improve.[49] China is dedicating itself to eliminating absolute poverty, reducing relative poverty, and narrowing the gap between rich and poor. It will never allow the emergence of city slums with high crime rates like those appearing in some emerging countries or underdeveloped inner-city communities in the United States. China's modernization is absolutely not designed to benefit just a portion of its people, cities, and regions. Rather, China's modernization aims to provide for the common prosperity of all people, across urban and rural areas and reaching both the coastal region and the vast interior hinterland. Such egalitarianism is the most significant difference between China's socialist modernization and the capitalist modernization program of the world's already developed countries. Despite the fact that the United States is the most developed country in the modern world, it still faces a plethora of social problems, including a wide gulf between the rich and the poor and pockets of extreme poverty. China aims to put its massive population of more than a billion people onto the path of prosperity, enabling them to attain a high level of human development, despite the fact that per capita GDP and resource consumption levels remain below those of developed countries.

China should also be a superpower with significant soft power resources. Chinese civilization must take its place within the diverse forest of global civilizations. There are many differences between China and the United States in terms of culture. American culture is exceedingly individualistic, with many concerned chiefly, or in some cases only, with themselves and their own advantages, to the exclusion of others. Chinese culture, on the other hand, is more tolerant and is guided by the principles of harmony, peace, and cooperation. These differences in culture will serve as the most prominent contrasts between the American and Chinese superpowers.

A few years ago, many people in the world were discussing the idea of a "Beijing consensus."[50] In fact, the Chinese discussion of world affairs does not stress consensus. Rather, Chinese scholars stress a type of Chinese opportunity, Chinese market, Chinese knowledge, Chinese culture, and a Chinese modernization path. The expression *Beijing consensus* implies that China is imposing its way on others. If anything, it should be known as the *Beijing proposal*. Other countries can choose whether or not they would like to accept it. In addition, they may accept it wholesale or accept it only in part. For developing countries, especially the least developed countries, China's attitude has always been to give more than it gets and to give before getting. For example, China was the first country to eliminate tariffs on goods imported from underdeveloped countries. China is also ready to incorporate the achievements of any civilization into its own culture, including freedom and democracy. Indeed, China promotes democratic change as a way of more effectively and more comprehensively increasing the welfare of its own citizens.

Finally, China must be a mature and responsible superpower. The hallmark of maturity is, domestically, a democratic, open, transparent, and well-structured system of government. Internationally, a responsible Chinese superpower should be cautious and not radical in its response to international issues, should ameliorate contradictions between peoples and countries rather than exacerbate them, and should encourage mutual trust rather than allow suspicions to grow. A Chinese superpower will be responsible not only for its billion-plus people but also for the billions of people all over the world. It must not only fulfill its commitment to provide national public goods but also take on the responsibility of providing public goods on a global scale.

China's reform and opening has become an acceptable political and ideological consensus in the country and will undoubtedly continue to serve as the driving force for future progress. By 2020 China is sure to exert greater influence on a more comprehensive set of issues across the globe. At the same time, it is likely to increase its contributions to mankind as a whole. China is looking forward to the development of the world, and the world is looking forward to the development of China.

Analytical Framework

The analysis employed in this book is based on multiple models. Discussions of China's rise are broad based, touching on many different issues. The narrative construction resembles the freehand brushwork of traditional Chinese painting rather than fine brushwork with close attention to details. The general analytical structure comes in the form of a progression from an explication of current conditions, to an analysis of these conditions, and then to a commentary on how these conditions are likely to change in the future.

The analysis requires looking at China's development from a scientific and rational perspective. For example, China's development must be examined from the two-dimensional perspective that takes into account both major per capita indicators and quantitative comparisons of other major indicators. This will provide insight into the dual character of China's development. In addition, any assessments of China must be both dynamic and holistic. Both the singular tree and the forest must be dissected, not only statically but also dynamically, so as to avoid one-sided views. Finally, China's political economy must be taken into account. In other words, one must analyze how political decisions respond to and influence economic development.

China's transition must also be considered. For example, it is necessary to understand why China's development entails so many challenges. In gaining such an understanding, it is important to note that Chinese development resonates closely with the challenge-and-response paradigm developed by the British historian Arnold Joseph Toynbee.[51] Chinese leaders consistently respond strongly to both internal and external challenges, harnessing the creative impulse necessary to respond to them. This is especially true in the economic domain. The Chinese leaders Deng Xiaoping and Chen Yun were at the forefront of reform. Their successors, Jiang Zemin and Hu Jintao, continue this heritage.

The question of how to interpret China's successes is also a salient one. Here, four interrelated innovations should be used as criteria of measurement:

—The theoretical innovations of Deng Xiaoping: Deng inspired Chinese citizens to "seek truth from facts" and to "emancipate their minds" in seeking social innovation.

—Institutional innovation: the transition from a planned economy to a market economy.

—Market innovation: the opening and development of the market.

—Technical innovation: the importation of new technology, innovations in domestic and foreign technology, and the promotion of independent technical innovation.

China's development must also be placed within the proper historical perspective rather than viewed simply within the context of the last thirty years of reform and opening. Both the first thirty years and the second thirty years of the PRC must be analyzed. In a way, Mao Zedong's mistakes made possible Deng Xiaoping's successes. The ten-year upheaval of the Cultural Revolution made reform and opening possible. It provided the circumstances necessary for the last thirty years of progress toward increased unity, stability, and prosperity.

Forecasts of China's future should be formulated using a dynamic mechanism of evaluation, not simply numbers. In their projections of China's development trajectory, many scholars underestimate the ability of the Chinese people and the Chinese government to learn and adapt. Simply looking at the transformation of China's rural areas in recent years can shed light on the significant progress China has made. Throughout my career I have participated in governmental policy-related debates, decisionmaking, and evaluation. In the course of this participation I have come to understand the difficulties inherent in forecasting China's future and have had the opportunity to learn from previous successes and failures. With this in mind, this book combines standard facts and figures with analysis of the complex inner workings of the overall process driving China's rise, providing the reader with a sophisticated way to predict the future of the PRC's development trajectory.

China has smoothly entered the post–Deng Xiaoping era. Unlike the generations of the Chinese leadership led by Mao Zedong and Deng Xiaoping, the post-Deng generations do not feature political strongmen. Instead, the replacement of political leaders has been institutionalized, standardized, and programmed, and the leaders themselves have become increasingly knowledgeable, professional, and open minded.[52] Through this process, political decisionmaking has become more scientific and democratic, and the government has broadened its focus to include public service generally rather than economic development alone. The higher order political ideal of "making the country strong and the people prosperous" continues to guide the government's decisionmaking process. The government also hopes to realize scientific development and, through such development, to improve people's lives, enhance national power, and increase international competitiveness so as to enlarge China's contributions to the world. These ideals form the starting point from which China's goals and strategies for the future are derived.

In sum, this book represents my efforts to observe China as an insider, to understand China as a researcher, to forecast China's future as a participant in its evolution, and as a scholar of the era following Deng Xiaoping, to help construct China.

Contents and Organization of the Book

Over the next decade China needs to mature from its current status as one of the world's economic powers into a comprehensive superpower. In assessing what China is likely to look like in 2020, this book addresses the following questions: How should one formulate predictions regarding China's rise? What kind of influence will China's status as a superpower have on its international behavior? What are the main goals China is striving to attain before the year 2020? And what national grand strategy should China adopt as it develops?

Chapter 2 is devoted to a discussion of China's economic development. In this chapter the historical trajectory of China's economic growth, which takes the form of a U curve, is explicated. Furthermore, 2020 is determined to be the likely year that the PRC will surpass the United States as the world's preeminent economic power. The chapter explores in greater detail how China has achieved its goal of catching up to the industrialized world economically and examines the sources of Chinese development. In doing so, the chapter emphasizes the ideological emancipation of the late 1970s, advocated by Deng Xiaoping, which significantly changed the mind-set of China's more than a billion citizens, unveiling a new era of creativity and productivity. Viewed from a macrolevel perspective, Chinese development is still nascent, with significant potential for future growth. As the chapter elucidates, China has manifested a "catch-up effect" and a "convergence effect" in both the international and domestic spheres. Finally, the chapter expounds on what trends and characteristics are likely to emerge in the course of China's future development.

Chapters 3 to 5 analyze China's social development and demographic makeup. The analysis begins with the understanding that the main risk to China's future growth lies in the disharmony between economic and social development. Economic growth has resulted in many social problems and social contradictions that have yet to be solved. One prominent example is the aging of China's population. In particular, chapter 3 examines population trends, including the demographic structure and labor force structure. Chapter 4 focuses on the daunting challenges that public health care will pose in the next decade. Through an examination of the attainment of a healthy China, the chapter highlights the importance of human development—the sum total of population and social development—which should supplement economic growth as one of the main objectives of modernization. It is imperative to increase standards of living in both urban and rural areas by 2020. Chapter 5 examines issues relating to human resources, especially the role of education as China transitions from a country that provides education in high quantities to one that provides education of both quantity and quality, making it an educational powerhouse in the world.

Chapter 6 discusses science, technology, and the state of innovation in China. The twenty-first century is defined primarily by the prominence of the knowledge economy. Technology and knowledge, which are based on innovation, will be the main sources of economic growth. In the process of developing its economy, China will do its utmost to catch up with other countries in terms of knowledge, will increase its number of professionals in science and technology, and will foster an environment conducive to innovation. Through these actions, China will gradually transition from a knowledge-closed nation into a knowledge-open nation and from a country lagging in science and technology into a country spearheading global innovation. The major science and technology development plans and initiatives adopted by China since the beginning of the twenty-first century have already opened up an innovative future.

Chapter 7 focuses on the challenges brought about by climate change and environmental degradation. The development model of fixing the environment after it is polluted is shortsighted, seeking instant benefits at the expense of the country's future ecological welfare. The result has been serious pollution and ecological destruction. China's vast swaths of highly varied land frequently fall victim to natural disasters. Despite impressive economic growth for more than two decades, and in part because of accompanying labor market transitions, millions of migrant workers in China were stranded during the unexpectedly intense snowstorms in January 2008.[53] With this in mind, it is imperative that the government prioritize protecting the environment and preventing, to the extent possible, natural disasters. Given that China is the world's top coal consumer, its top emitter of sulfur dioxide, its second-largest energy consumer, its largest carbon dioxide emitter, and one of its worst victims of climate change, it is also extremely important for the PRC to find a way to combat global warming.

Chapter 8, the concluding chapter, discusses China's grand strategy and 2020 development goals. It has been the aspiration of several generations of

Chinese leaders to construct a strong country populated by economically well-off citizens. This aspiration will be realized by 2020. By then, China will have the world's largest economy, will be the global leader in innovation, will be the largest environmentally friendly society, and will serve as a new base for contemporary world civilization. These goals can be realized only by relying on a grand strategy formulated with China's unique national conditions in mind. Such a strategy is detailed in chapter 8 and includes the following facets: human development, knowledge-based economic development, green development, peaceful development, and a comprehensive global strategy.

If current development trends continue, one can expect that the day when China overtakes the United States in a variety of respects—not only in economic power but also in human capital and science and technology capacities—is not far off. It is important that as China prepares for this day, it take on more responsibilities and make greater contributions to international society not only in terms of economic development but also in terms of culture, science, technology, and ecology.

CHAPTER TWO

ECONOMIC DEVELOPMENT: PAST, PRESENT, AND FUTURE

The rise of China has become a hot topic of discussion across the globe. China's huge population and labor force have ensured that its economic development evolves differently from the experience of other countries. At present there are four nations with a labor force in excess of 100 million people: China (786 million), India (448 million), the United States (157 million), and Indonesia (111 million).[1] In only two countries does the number of scientists and engineers engaged in research and development (R&D) exceed 1 million: China (1.82 million) and the United States (1.4 million).[2]

One economic implication of China's rise is that the proportion of the world's total GDP accounted for by China will increase; at the same time, its GDP, its imports and exports, and its domestic consumption will approach U.S. levels. As the most populous developing country, China will, by 2020, overtake the United States, to become the world's preeminent superpower. China's ascent to such prominence will not only demand worldwide attention but will also influence human development worldwide.

Britain's Industrial Revolution was termed a revolution because, for the first time, it was possible for average per capita incomes to double within a few generations. In the heyday of U.S. development, incomes more than doubled in a single generation.[3] During the thirty years between 1978 and 2009, China's GDP grew 18.6 times, averaging annual growth rates of 9.9 percent. In addition, China's per capita GDP grew 13.1 times, growing at an average rate of 8.7 percent a year.[4] Put another way, per capita income doubled in just eight years. Despite having a much larger population than either Britain or the United States had during their respective periods of rapid development, China was able to achieve similar, if not better, results. Thus the PRC's growth should be considered the Chinese economic revolution.

The outcome of this revolution is that the PRC has risen swiftly from the most populous and destitute nation in the world to an economic power ranked among the world's elite in terms of aggregate GDP. Using market exchange rate measures to compute GDP, China's economy was the world's tenth largest in 1978, eleventh largest in 1990, sixth largest in 2000, fourth largest in 2005, and third largest in 2009, trailing only the United States and Japan.[5] Furthermore, China has become the world's second-biggest economy, according to data released on August 16, 2010—Japan's economy having fallen behind China's in the second quarter of 2010.[6] China's economy, measured by PPP, was the fifth largest in 1980, following the economies of the United States, the Soviet Union, Japan, and Germany. By 1992 the Chinese economy had overtaken those of Germany, the Russian Confederation, and Japan, becoming the world's second-largest economic power based on the PPP method, a position it has maintained to the present day. Over that same period of time, the gap between China's GDP and that of the United States narrowed by a factor of 4.37 in 1978, 1.86 in 2000, and 1.06 in 2008. It is estimated that China will overtake the United States at or around the year 2020—or even earlier.[7]

This chapter addresses the following questions: What is the history of China's economic growth? How has China managed to catch up so quickly in the domains of industrialization and modernization? How will China overtake the United States, becoming the world's biggest economy? What impact will China's rise have on the rest of the world? What impact will the global financial crisis have on China's economy? What is the likely development trajectory for China's economy?

The U Curve: The Historical Trajectory of China's Economic Growth

The history of world modernization is equivalent to a race to industrialize. The evolution of world economic development shows that, through the process of modernization, many countries that began the process of industrialization late have since caught up to, or have overtaken, their forerunners. Prominent examples of this phenomenon include the United States, Japan, Asia's four Tigers (Hong Kong, Singapore, South Korea, and Taiwan), and now the PRC.

The United States began its economic takeoff in 1870, five years after the conclusion of the Civil War. Through average GDP growth rates of 3.94

percent a year over the course of forty-three years, the United States was able to increase its economy from 76.6 percent of Britain's in 1870 to 107.7 percent of Britain's by 1913. The rapid growth of the U.S. immigrant population accelerated this process. In 1830 the British economic aggregate was 2.3 times that of the United States, and the British population was 1.8 times that of the United States. By 1913, however, the U.S. economic aggregate was 2.3 times that of Britain, and its population was 2.1 times greater.[8] Through this growth the United States became the world's leader in economic power and innovation, a status that it has maintained through to the present day.

Japan began its economic takeoff in 1950, following the conclusion of World War II. From 1950 to 1973, the year of the first oil crisis, Japan was able to maintain an annual GDP growth rate of 9.29 percent. Over the course of that period, Japan's per capita GDP gap with the United States decreased by a factor of 5.0 to a factor of 1.5.[9] The rapidity and success of Japan's growth led many to regard it as the world's greatest economic success story to date.

The four Asian Tigers began their rise in the 1960s, taking the next thirty years to catch up with other developed countries.

China's economic takeoff began in 1978, when it implemented reform and opening in order to narrow the gap between its economy and the economies of the United States, Japan, Western Europe, and the Asian Tigers. Looked at in terms of population size and enhancement of human capital, China's economic rise very much resembles that of the United States. A country can never become a strong global economic power without a sizable population. When China began its economic takeoff its population stood at 960 million. If one looks at China's economic rise in terms of speed of growth, China is more like Japan. Japan, however, maintained its 9 percent growth for only twenty-three years, while China has already maintained commensurate rates for more than thirty years. Viewed from a technical perspective, China's rise is more like that of Asia's Tigers, which used export-oriented growth strategies and foreign direct investment to drive development. The principal difference between China and the Asian Tigers is that the latter group have small populations and small domestic markets, while the PRC has a huge population and a sizable domestic market. All of these facts—the size of China's population, its reform and opening policy, and its long-term sociopolitical stability—mark the PRC as a unique country. And its economic rise is certainly unparalleled.

From a historical perspective, China is indeed a special case. China is, and has almost always been, a political unit larger than any other.[10] From the year 1 AD through to the beginning of the nineteenth century, China, despite its largely agrarian economy, featured both a large population and the world's most powerful economy. In the tenth century China led the world in per capita income, a position it maintained until the fifteenth century. Throughout this period China outperformed Europe in terms of technological development, the intensity with which it used its natural resources, and its capacity for administering an extensive territorial empire.[11] Between 1500 and 1820, during which China was the world's most populous country with the world's largest economy, China's agricultural civilization peaked. Toward the end of this period, however, per capita GDP growth stagnated, at first slowing down and then assuming a downward trend, an indication that economic growth had been driven mainly by the intensive input of labor rather than productivity.

After 1750 the Western world, inspired by the invention of the steam engine, launched the first Industrial Revolution. In Western Europe scientific discoveries provided the basis for new technologies that drove its economic growth. In the middle of the nineteenth century the second Industrial Revolution, characterized by the spread of railways and electricity, swept across sections of Europe, North America, and Oceania, transforming agricultural civilizations into industrial societies. China remained untouched by these waves of modernization, maintaining an agricultural society. Compared to the industrializing countries, China lacked the necessary innovative capacity to modernize and thus entered a period of economic decline. While the historical trajectory of economic development in the industrialized countries can be modeled using an inverted U curve, China's development path can be modeled using a standard U curve.

Between 1820 and 1950 China's GDP as a percentage of the world's total plummeted from one-third to less than one-twentieth. Its autarkic status led to development levels far behind those of the industrialized world, weakening China to the point where it could be subjugated by imperial powers.[12] As of 1950 China, the most populous developing country in the world, also had the largest number of people subsisting on less than a dollar a day (540 million people, or 40 percent of the world total).[13] In 1950 China's per capita GDP was only a fifth of the world's per capita GDP, or less than a tenth that of the twelve Western European countries.[14] When China first launched its process of industrialization, modern industries accounted for only between one-tenth and one-seventh of total GNP.[15] The mortality rate of the Chinese population was 20 per thousand, the infant mortality rate was as high as 200 per thousand, and mean life expectancy at birth was only thirty-five years.[16] The average number of years of education that individuals aged fifteen or older had received was approximately one.

China in 1950 was a country in dire straits. It was one of the poorest countries in the world both in terms of economic and social indicators.[17] Indeed, it had the world's lowest human development indicator at the time—0.225, slightly lower than India's 0.247.[18] Against this backdrop of extreme backwardness and poverty, China launched its ambitious programs of industrialization, urbanization, and modernization.

After 1950 China entered a period of economic growth, with per capita GDP beginning to grow at over 1.0 percent. Despite the challenges to Chinese development posed by the Great Leap Forward and the Cultural Revolution, China realized the goals it set for industrialization in the 1950s and 1960s. For instance, it rapidly attained an elementary level of state industrialization and established relatively complete national industrial and economic systems, paving the way for future development. China also succeeded in feeding one-fifth of the world's population with only 7 percent of the world's arable land and 6.5 percent of its water. China's pre-1978 social and economic development cannot be underestimated.

By 1978 the average number of years of education received by people age fifteen or older exceeded four years, the adult illiteracy rate had fallen to 33 percent (from 80 percent in 1949), and China had become the world's leader in total human capital by number (human capital is equivalent to the number of people in the age group fifteen to sixty-four years multiplied by the average years of education received by people above the age of fifteen).[19] In addition, the health of both urban and rural citizens had improved greatly, with the mean life expectancy at birth improving and the mortality rate of both infants and pregnant women dropping significantly. One good measure of China's progress over this period is the dramatic increase in its human development indicator to 0.523 (1975 figure).[20] Most important, China had created a more equitable society than any other in modern history. Within such a society, the people, especially workers and farmers, attained the status of masters of their country.[21] At the same time, China became a world power in terms of comprehensive national strength—economic, political, and military—cementing its heightened status in the international community. According to my calculations, China rose to the rank of fourth in comprehensive national power by 1980, or 4.7 percent of the world total, just after the United States (22.5 percent), the former Soviet Union (figures unavailable), and Japan (6.0 percent), but still higher than India (3.4 percent).[22]

In the early days of the PRC, China replicated the Soviet economic growth model, putting great emphasis on the development of heavy industry. Later, Mao Zedong tried to break the "dogmatism" of the Soviet model by developing a distinctly Chinese model, but failed. As of 1978 China was at once a world economic power and one of the globe's poorest countries, with extremely low per capita gross national income levels.[23] At the time most Chinese people were deemed to be poor or within the low-income strata. According to the World Bank's definition of absolute poverty—subsisting on less than a dollar a day—the impoverished in China numbered 490 million. By 2002 this number would drop to 88 million, 6.9 percent of the population, rather than 49 percent of the population (1981 figure).[24]

It is important to note that Deng Xiaoping's economic reforms—which were designed to promote economic growth, to reduce the number of people living in poverty, and to secure an all-around well-off society—were entirely different from those instituted by the former Soviet Union before its dissolution. The results also are very different. As Angus Maddison states, after a period of reform, China rose again while the Soviet Union collapsed.[25] In 1978 China's GDP was less than that of the Soviet Union. As of 1990 China's GDP was 1.85 times the size of Russia's. By 2006 China's GDP was 7.1 times greater. In 1978 China's per capita GDP was only 16 percent that of the Soviet Union. By 2006 it had reached 89 percent of that of Russia, the largest of the Soviet successor states. The vastly different results of the Russian and Chinese reforms are demonstrative of the critical importance of choosing the right reform strategies and paths.

Sources of Economic Growth in the PRC

China's economic growth has two main sources: the input of the factors of production, such as land, labor, capital, energy, and other resources; and total factor productivity (TFP), or the productivity of the allocation and use of the factors of production. Increases in physical capital have a positive effect on welfare. Human capital and natural capital not only affect welfare through growth but also are an important part of the concept of welfare itself. In addition, human capital and natural capital can contribute to the accumulation of physical capital by increasing the returns on physical capital. Similarly, the accumulation and effective utilization of physical capital can increase returns on the use of both human and natural capital. Unlike tangible production factors, TFP is intangible and can be divided into structural factors and knowledge factors. The structural effect refers to a factor's ability to advance the movement of other factors, raising the efficiency level in allocation of resources. Examples of the importance of the structural effect can be found in the movement of agricultural labor to nonagricultural activities, the movement of labor from low-efficiency sectors of the economy to high-efficiency sectors, the expansion of economic scale and market size, the effect of specialized divisions of labor (domestic and international), and the expansion of internal and external trade. The knowledge effect refers to knowledge and technology factors that have a decisive long-term impact on economic growth. These include large-scale technology imports, constant increases in the quality of labor, investment in human capital, development of information and telecommunication technologies, absorption of external knowledge, increased investments in research and design, and increased capacity for domestic innovation.

The engine and sources of China's economic growth can be divided into two stages. The first stage is the planned economy stage (1952–77). During this period, average annual GDP growth was 6.1 percent, with GDP expansion in 1952–57 averaging 9.2 percent a year. In my previous books I refer to the period between 1952 and 1957 as the first golden age of modern Chinese economic growth.[26] Looking at this period from the perspective of long-term development, it was both an indicator of China's high economic growth potential and an important hallmark in the country's later economic takeoff. Although economic growth tapered off in the twenty years following this first golden age, it would return to, and often exceed, 1952–57 levels after the institution of reform and opening in 1978. The causes of the twenty-year gap between China's first golden age of economic growth and current high rates of GDP expansion were the Great Leap Forward and the Cultural Revolution.[27]

When Mao Zedong visited Moscow in November 1957, Soviet premier Nikita Khrushchev advanced the idea of overtaking the United States in terms of total and per capita output of manufactured goods within fifteen years. Strongly influenced by Khrushchev, Mao Zedong promoted a similar policy, announcing that China would overtake Britain in output of iron, steel, and other major manufactured goods within fifteen years. The first-generation Chinese Communist Party (CCP) leader Liu Shaoqi announced the policy at the end of 1957. Only a year later, Mao, feeling that China's growth was not proceeding rapidly enough, launched the Great Leap Forward. The stated goal of the GLF movement was to overtake Britain in iron and steel output within just two years, overtake the Soviet Union within four years, and overtake the United States within ten years.[28] Despite the fact that economic growth rates for 1958 were reported to be 21.3 percent, the GLF led to an unprecedented economic crisis. In 1961 GDP growth stood at negative 27.3 percent, climbing to a still anemic negative 5.6 percent by 1962. By 1965 GDP was 41 percent less than it would have been had the Great Leap Forward not taken place.[29]

After a brief period of recovery in the early 1960s, Mao launched the Cultural Revolution in 1966. This political movement also had an adverse effect on economic growth, leading to GDP expansion rates that were significantly lower than China's 9 percent long-term potential growth rate.[30]

In terms of the sources of growth, China mainly relied on capital input, which was responsible for 75 percent of the economic growth from 1952 to 1977. Human capital and labor inputs were responsible for 20 percent and 13 percent of economic growth, respectively. TFP over the course of this period was negative, preventing the country from achieving sustained growth. These numbers indicate that the most prominent features of China's planned economy era were high capital inputs and factor inputs but low economic efficiency and TFP.

The second stage is the reform-and-opening stage (1978 to the present). The rate of economic growth during this period was higher than that of the planned economy period primarily because TFP increased significantly, with its contributions to economic development rising steadily. From 1978 to 2005, capital input growth dropped 2 percentage points but was still responsible for about 39 percent of economic growth. Human capital growth also dropped significantly, constituting only about 7 percent of economic growth. Labor input also dropped, albeit only slightly, making up about 8 percent of economic growth. TFP, however, grew greatly, reaching 4.4 percent and contributing the most to economic growth (46 percent; table 2-1).[31]

Why was it that TFP grew from negative to positive after reform and opening? A purely economic explanation is not sufficient. Answering this question requires adopting a new perspective.

The Cultural Revolution launched by Mao Zedong, or Mao's blunder in his late years, is the mother of the success of Deng Xiaoping's economic reform.[32] China's model of reform differs from those of other countries and cannot be interpreted using existing economic growth theories or models. The Chinese refer to their model of reform as the mind emancipation model. I refer to it as mind emancipation and concept innovation–based economic reform. This is, perhaps, the most salient feature of China's reform. It may also be the most significant of China's modern contributions to the world.

The mind emancipation and conceptual changes encouraged by Deng Xiaoping are like idea-based growth models, in which ideas (such as knowledge, experience, and innovation) are inputs that produce progressive increases in marginal returns. For the time being I will call the idea input an intangible factor input (or soft input) to differentiate it from such tangible factors as capital, labor, and resources (or hard inputs). Even if hard inputs are not

Table 2-1. *Sources of Economic Growth, China, Two Periods,*
1952–77 and 1978–2005

Percent

Source	1952–77	Contribution	1978–2005	Contribution
GDP	6.1	100	9.6	100
Capital stock, 1987	11.5	75	9.5	39
Employment	2.6	13	2.4	8
Human capital	4.1	20	2.1	7
Total factor productivity[a]	–0.5	–8	4.4	46

Source: Hu Angang and Liu Taoxiong, "Comparison of Defense Capital Power among China, U.S., Japan, and India," *Strategy and Management* 6 (2003); Chang Li, "Study of Regional Gaps in Human Capital and Economic Development" (master's thesis, Tsinghua University); *Chinese Labor Statistical Yearbook 2000; China Statistical Abstract 2004*, pp. 38–42.

a. Total factor productivity calculated using capital input (0.4), labor input (0.3), and human capital input (0.3).

increased, soft inputs can still stimulate economic growth. Ideas have another special effect in China. As they spread across the country, they serve as a catalyst for changing the thinking of its 1.3 billion people, thus releasing their capacity to be both productive and creative.

Mao Zedong advanced this idea as early as 1963, though his words did not amount to an economic model. He said: "Once the correct ideas characteristic of the advanced class are grasped by the masses, these ideas turn into a material force, which changes society and changes the world."[33] This was Mao's theory of the transformation of spiritual wealth into material wealth. From this one can see that China's reform and opening was not imposed from the outside but rather originated within China. It represents a movement of self-reform, self-improvement, and self-development. Reform and opening was advanced by Chinese leaders in response to both internal and external challenges and was launched and promoted with vigor.

China's reform and opening is quite different from the reforms undertaken by Eastern European countries and the former Soviet Union. China adopted the method of changing people's thoughts instead of "replacing the blood of the whole body." Chinese leaders reached a political consensus and launched a movement emancipating the minds of individuals both inside and outside the party, preparing the public for the advent of reform and opening. At a June 1978 political work conference of the People's Liberation Army, Deng Xiaoping called on the party leadership to shatter the mental shackles that had limited its leaders' ability to innovate. This approach to reform is quite different from those of Eastern European countries and the former Soviet Union, which adopted major changes in leadership and new political parties as part of their reform packages. Their approach replaced monolithic communist parties with groups led by new "rebels" who wanted to "build Rome overnight." Russian GDP growth after 1990 assumed the shape of a U curve. Its GDP in 1996 was only 59 percent of its 1990 level. As of 2008 Russian GDP was 117 percent of its 1990 level. Over a period of sixteen years Russia stagnated while China continued to develop. In 1990 China's GDP was 1.85 times that of Russia. By 2008 it was 6.95 times greater.[34]

The communiqué of the Third Plenum of the Eleventh Party Central Committee was issued in December 1978, initiating a new round of collective learning and mind emancipation, which shattered the shackles of dogmatism that had plagued the CCP. Hu Qiaomu was the earliest to rise up against the dogmatism of Mao's theory of continuous revolution under the dictatorship of the proletariat. Much of this reorienting occurred at the beginning of January 1979, just after the end of the Third Plenum. The bold theoretical breakthrough brought the party back to the resolutions of the Eighth National Party Congress in 1956.

In spite of these necessary ideological revisions, the basic idea behind the concept of mind emancipation has its roots in Mao Zedong's thought of seeking truth from facts. The principle of seeking truth from facts does not advocate using books as a guide to life but rather required leaders to use the study of new problems and new circumstances to craft effective solutions. Mao's thought and language may not have been the best in the world, but they were and remain the most suitable for China as well as the easiest for the Chinese people to accept. From Mao's ideas, Deng Xiaoping derived many conceptual innovations concerning reform and opening.

The first of Deng's conceptual innovations is widely known as the cat theory. It holds simply that it doesn't matter if the cat is black or white as long as it catches mice. It was first advanced by Marshal Liu Bocheng and later revived by Deng Xiaoping when he was instituting reform in China's rural areas during recovery from the Great Leap Forward.

The second conceptual innovation is the truth and fact theory, advanced by the prominent economic thinker and official Chen Yun. He held that "there should be no blind obedience to superiors or books; there should be obedience to truth and facts only; there should be exchange, comparison, and repetition." This theory is a summary of Chen Yun's life story, one defined by revolution and economic construction. His most important ideas were those that challenged the dogmatism of the CCP.

The third theory is the theory of exploration, also known as the theory of crossing the river by feeling the stones, or the theory of feeling our way forward. This idea, too, was advanced first by Chen Yun. In December 1980 Chen stated the following: "We must carry out reforms, but the steps must be steady, because we shall encounter many complicated problems. So do not rush. . . . We should proceed with experiments, review our experience from time to time, and correct mistakes whenever we discover them, so that minor mistakes will not grow into major ones. This is 'to feel out the way forward.'"[35] Deng Xiaoping later expressed these very same ideas.[36]

If viewed from the perspective of information economics, this exploration theory is equivalent to the trial and error, or optimum search, method of economic management. Seeing as the reformers were faced with many uncertainties, feeling the way forward was in fact a way to find a bridge or a boat that would take China from one side of the river—poverty—to the other—wealth. Many countries have failed in their quest to reform because they failed to find a bridge or boat that could shepherd them across the turbulent river of economic transformation. By feeling the way forward, China could prevent significant economic fluctuations. There are many postulations in economic theory, but China's reform is a reality that has confounded many of them. At this point it would be impossible to reject all of the previous moves and start anew. The only way out is the trial-and-error method, which seeks not only the maximization of gains from reform but also the minimization of risks.

The ability of Chinese leaders to learn increased markedly during the reform era. Chen Yun, Li Xiannian, and Bo Yibo were the designers of China's planned economy during the Mao Zedong era.[37] When Deng Xiaoping came to power, however, they became the designers and promoters of institutional reform. The reform of China was a classic case of a transformation orchestrated by party insiders. These CCP leaders, fully aware that the planned economy did not suit China's national conditions, moved to implement reforms. Able to turn the historical lessons and memories of the internal upheaval of the Cultural Revolution into capital for moving China forward, Deng Xiaoping and Chen Yun managed to avoid the disintegration of the Soviet Union and open up a new path for China.

China's Potential for Further Economic Growth

China is still in the takeoff stage of its economic development and is likely to maintain annual GDP expansion rates of 8–9 percent for quite some time. The sources of China's future economic growth will be the impact of the international and domestic catch-up effect, high domestic savings and investment rates, significant increases in human capital, the growth of the nonagricultural sector, and steady increases in total factor productivity.

INTERNATIONAL AND DOMESTIC CATCH-UP EFFECT

Because China's per capita GDP is still low it is likely to experience two types of catch-up effect. The first is the international catch-up effect. Generally speaking, the bigger the gap in per capita GDP between a low-income country and a high-income country, the higher the developing country's economic growth rate. China has taken full advantage of this catch-up effect since 1978. According to calculations by Angus Maddison, China's 2005 per capita GDP (in 1990 international dollars) was 18.3 percent of the U.S. total, 57.5 percent of South Korea's, and 62.4 percent of Taiwan's.[38] World Bank figures show that China's 2005 per capita GDP (2005 international dollars) was 9.8 percent of the U.S. figure, 51.2 percent of South Korea's, and 62.6 percent of Taiwan's. Since the 2005 per capita GDP gap between China and the United States is still larger than the 1975 per capita GDP gap between South Korea and the United States, this significant gap will likely allow the PRC to continue to grow at a high rate for another generation (2005–30).

The second type of catch-up effect is internal. Domestically, there are large disparities among China's regions. The per capita GDP of the central and western parts of the country is significantly lower than that of coastal areas. Thus China's underdeveloped zones are likely to experience their own domestic catch-up effect. Since the beginning of the new millennium, GDP growth rates in China's interior have picked up speed. In addition, seven provinces and autonomous regions in the central and western regions have experienced decreases in total population, as people migrate to the coastal areas. Beginning in 2003 a trend of regional convergence in per capita GDP appeared, clearly indicating the existence of a domestic catch-up effect. As a vast country with great regional disparities, China has great potential for development. Shanghai, Beijing, and other coastal cities have already narrowed the per capita GDP chasm between themselves and developed countries. It is also likely that most of the central and western parts of the country, whose per capita GDP remains far behind the highest-income countries, will experience a long-term period of rapid economic growth. The World Bank data indicate that even the slowest-developing province, Qinghai, had an average per capita GDP growth rate of 6.9 percent from the mid-1980s to late 1990s, which exceeded South Korea's 5.8 percent. Indeed, Qinghai's per capita GDP has registered growth in excess of 10 percent for seven years running (2001–08).[39]

It is important to note that the two types of catch-up effect are not guaranteed to occur. They require two kinds of opening up. The international catch-up effect, for example, will not be experienced by a country unless that country is participating in economic globalization, integrating into the world economy, and actively communicating with other countries (especially highly developed countries). To experience the internal catch-up effect, a country must create a large, unified domestic market and take steps to stimulate domestic market competition. It must also attempt to integrate its intracountry regional economies.

HIGH DOMESTIC SAVINGS AND HIGH INVESTMENT RATES

Rising Chinese domestic savings and investment rates are likely to fuel future economic growth. While China's domestic savings rate grew from 43 to 55 percent between 1995 and 2007, the U.S. savings rate dropped from 16 to 14 percent.[40] Zhou Xiaochuan, the governor of the People's Bank of China, believes that China's rapid rise in domestic savings is due to both the steady increase of savings by residents and significant increases in savings by businesses.[41] Viewed from the perspective of the Italian American economist Franco Modigliani's life-cycle hypothesis, the rise of the non-labor-force population as a proportion of the total population encourages individuals to save money for old age, thus leading to the rise of personal savings rates. Viewed from a perspective that focuses on different economic growth stages, the extraordinary rapidity of domestic GDP growth has resulted in individuals saving most of their increases in income, leading to the higher than normal savings rate. Both theories are valid in China's case, making it easy to understand.[42]

Also of note is that the gross capital formation (GCF) of China as a proportion of GDP has assumed an upward trend. During the Ninth Five-Year Plan (1996–2000), the GCF made up 36.5 percent of GDP.[43] Over the course of the next five-year plan, that proportion rose to 40.7 percent, climbing to 42.8 percent between 2006 and 2008.[44] World Bank data show that China's GCF averaged an annual growth rate of 11.5–13.4 percent between 2000 and 2007. During those same periods, the U.S. GCF averaged 7.5 percent and 2.5 percent.[45] Wang Xiaolu estimates that between 2006 and 2020 China's physical capital input would grow 13 percent.[46] I believe that it will grow by roughly 10 percent. Regardless, it will undoubtedly serve as the primary source of China's sustained high economic growth. The relatively low growth of labor inputs, coupled with per capita capital growth in excess of 9.7 percent, has led to an accelerated growth in labor productivity. In fact, with the rise of domestic investment rates, labor productivity has risen steadily, averaging

7.6 percent a year between 1995 and 2000 and 9.4 percent a year between 2001 and 2009.[47] Rapid increases in China's labor productivity represent the country's most important competitive edge in the international market.

EXPECTED INCREASE IN HUMAN CAPITAL

China has maintained a fairly high rate of human capital growth. Since the beginning of the twenty-first century, China's human capital growth rate has risen even quicker than before. From 2000 to 2008 the number of people with a high school education increased from 140.68 million to 171.31 million, averaging an increase of 2.49 percent a year. Over the same period the number of people with university educations increased at a more rapid rate, 7.87 percent, or from 45.63 million to 83.67 million.[48] By 2020 that number may reach nearly 200 million, more than the total number of laborers in the United States. In 2009 the number of individuals working on R&D in China reached 3.18 million.[49] In 2008 there were 1.61 million scientists and engineers.[50] This far outpaces the 1.4 million scientists and engineers in the United States in 2005. China is working to implement a medium- and long-term educational reform and development program, along with a medium- and long-term personnel training program. The core objective of both initiatives is to ensure that China is a powerhouse in the realm of human resources and personnel by 2020.[51] In fact, China's human resources situation constitutes the most important strategic advantage it has. From 2006 to 2020 human capital input, or the average years of education received by people above the age of fifteen, will grow an estimated 2 percent. In addition to human capital's direct contributions to economic growth, increases in human resources are also likely to indirectly contribute to development.[52]

GROWTH OF THE NONAGRICULTURAL SECTOR

While China's labor input growth has slowed, employment levels in the non-agricultural sector have soared. With the entrance of China's working-age population into a period of zero growth (< 0.2 percent), labor input growth has shrunk and is likely to hover at around 0.3 percent annually between 2006 and 2020, contributing little to economic development. At the same time, a large number of agricultural laborers have transitioned to nonagricultural forms of work, contributing to increases in labor productivity. Between 2002 and 2009 the number of agricultural laborers in China decreased by 71.62 million, to 297.08 million.[53] By 2020 China's agricultural labor force will fall to 200 million. Between 2000 and 2009 the number of laborers in the nonagricultural sector increased 3.3 percent annually, from 360.42 million to

482.87 million.[54] In the future, the average annual growth of labor in nonagricultural sectors may remain around 3 percent, which translates to a very high labor input in the nonagricultural sectors.

STEADY INCREASES IN TOTAL FACTOR PRODUCTIVITY

China's TFP, the key factor to its high economic growth, has continually increased in recent years.[55] It is estimated that between 2006 and 2020 China's GDP, bolstered by an ever-increasing TFP, will grow at an average annual rate of 8–9 percent. If TFP can maintain the 4.4 percent growth rate that it has averaged over the last thirty years, we can calculate (by employing the Cobb-Douglas production function) that GDP expansion is likely to top 8.5 percent annually, perhaps even reaching 9.0 percent. For this to occur, China must maintain a sustainable, stable, and healthy trajectory of economic development.

As China's rapid economic growth continues, the country's economic structure will adapt and evolve. An increasingly diversified economic structure and major adjustments to the industrial structure directly impact interest structures. The economy will become more and more balanced, especially in its imports and exports and in its levels of savings versus consumption; its balance of payments can be changed from a surplus to a deficit plus greater domestic consumption. The interests of the country will be affected by the above changes, and public policy circumstances will be transformed to support more balanced and sustainable economic development.

With the growth of per capita income, China's urban and rural consumption structures are being upgraded and diversified. Engel coefficients indicate that the economic conditions in both cities and towns are improving from a well-off (*xiaokang*) level to a better-off (*fuyu*) level—that is, from a society of modest means to a society of middle-level income. The proportion of income spent on food consumed at home has been dropping steadily, with expenditures on dining out increasing. These developments have stimulated the development of food processing and catering services. Spending on education and cultural and recreational activities ranks second, followed by spending on transportation, telecommunications, and housing. The use of mobile phones is nearly universal, and computer use continues to grow at an ever accelerating rate.[56] In cities, car ownership numbers are skyrocketing, accompanied by an increase in the demand for related services.[57] This has provided the microlevel basis for stimulating the development of services, especially modern services, such as financial services, high-tech services, and so on.

China's Engel coefficient over the last sixteen years, however, indicates that rural living standards lag behind urban living standards. Rural standards of

Table 2-2. *Engel Coefficients and Living Standards, Urban and Rural China, Selected Years, 1978–2020*
Unit as indicated

Year	Urban population (100 million)	Engel coefficient (percent)	Living conditions	Rural population (100 million)	Engel coefficient (percent)	Living conditions
1978	1.72	57.5	Adequate food and clothing	7.9	67.7	Absolute poverty
1990	3.02	54.2	Adequate food and clothing	8.41	58.8	Adequate food and clothing
2000	4.59	39.4	Well-off	8.08	49.1	Well-off
2009	6.22	36.5	Well-off	7.13	41.0	Well-off
2015	7.11	30–32	Better-off	n.a.	35.0	Better-off
2020	7.94	22–25	Better-off	n.a.	30–32	Better-off

Source: *China Statistical Abstract 2010*, pp. 39, 116; figures for 2015–20: author's calculations.

living in 1990 were equivalent to those of China's cities in 1978. In 2000 this twelve-year gap was narrowed to five years, with rural living standards roughly equal to 1995 urban living standards. By 2006, however, the gap had widened to seven years, with rural standards of living equal to those found in China's cities in 1999 (table 2-2). Changes to China's rural and urban Engel coefficients over the last three decades show that China transitioned first from a state of absolute poverty to one where its citizens had enough food and clothing then to a well-off society before finally arriving at better-off living standards. Overall, the living standards of China's more than a billion people have improved. However, development gaps between urban and rural areas have continued to widen. Differing paces of development ensure that although both urban and rural areas are experiencing a constant drop in the Engel coefficient and are transitioning from well-off to better-off living standards, there remains a chasm between their levels of modernization.

China is also undergoing significant changes in its production structure. Agricultural production as a proportion of GDP has continued to drop, from 14.8 percent in 2000 to 10.3 percent in 2009. Development trends indicate that this percentage will likely continue to drop, falling to an estimated 7.0 percent by 2015. On the other hand, industrial production as a proportion of GDP has risen, reaching 39.7 percent in 2009, only 0.4 percentage points higher than the 2002 figure.[58] This percentage places China within the ranks of the highly industrialized or the transitioning industrialized countries. This proportion is likely to continue to increase in the future before leveling off.[59] China failed to achieve the goals of its Tenth and Eleventh Five-Year Plans for expanding the service industry by 3 percent annually. In fact, the ratio of the service sector in overall GDP was even smaller than in 2001: by the end of 2008 it had decreased by 0.1 percent. Looking forward, the proportion of China's GDP composed of the service industry, especially modern services, should increase. The services sector must become a major contributor to economic growth.

Between now and 2020 China will transition from a low middle-income country into a middle-income country, thus becoming increasingly better off, with at least 800 million people enjoying higher standards of living. Given that China is still in the economic takeoff stage, it must maintain high savings and investment rates along with high levels of human capital input. In general, labor input growth will likely drop, while the number of high-quality skilled laborers increases, pushing up the cost of production factors. Even so, China will still maintain high growth rates in TFP, which translates into great potential for future economic development.[60]

Thirty years of experience implementing the policies of reform and opening have led the Chinese government to become increasingly mature in stabilizing the macroeconomy, making China highly resistant to negative external economic forces. This was evident in China's strong economic performance and quick recovery in the wake of the global financial crisis. Because national regulatory tools have become increasingly diverse and flexible, the economic fluctuation coefficient will likely continue to decrease. China's overall economic structure, consumption structure, production structure, employment structure, and trade structure will continue to change rapidly. Currently the "world's factory," the PRC will become a world service provider and a center for research and development. Most important, it will take full advantage of its huge domestic demand by becoming an international market.[61]

How Will China Catch Up with the United States in Total GDP?

China's position in the world in terms of economic aggregate (proportion of the world's total GDP) will continue its upward trend.[62] If calculated by the market exchange rate, China is already the second-largest economy in the world in 2010; if calculated by the PPP method, China was the second-largest economy in the world in 1992.[63] The next objective of China's economic growth is to catch up with, and then overtake, the United States in terms of aggregate GDP. There may be no way to accurately calculate exactly when

China's GDP will surpass that of the United States, but it is only a matter of time. In the opinion of some, this is the first time that the United States has genuinely experienced this sort of competitive economic challenge (unlike the exaggerated U.S. fears of Soviet economic might in the 1950s and of Japan in the 1980s).[64] These developments have even led experts from the U.S. government's Congressional Research Service to begin studying when China could catch up with and overtake America.[65]

How long will it take for China to catch up with—or overtake—the United States in GDP? There is no concrete answer to this question. The twentieth-century invention of the system of national accounts (SNA) made possible the computation of wealth created by any country.[66] Even so, the information is limited, uncertain, and diverse. Thus it is impossible to ensure the completeness, accuracy, and reliability of all information. This is referred to as the GDP uncertainty principle, an idea conceptually similar to the uncertainty principle in quantum mechanics.[67] Despite uncertainty, one can speculate on when China is likely to become the world's greatest economic power. Such speculation may not be perfectly accurate, but it can provide a certain measure of guidance, with the reality likely to be between the conservative and the more optimistic predictions. The following are three ways to measure China's catch-up with the United States.

MEASURE 1, PHYSICAL QUANTITY

The quantity method uses certain physical outputs, such as agricultural production or steel production, to compare two or more countries' economic power.

Both China and the United States are powerful in terms of agricultural resources. According to the UN Food and Agriculture Organization, total arable land in both countries in 2000 was considerable: the United States, 176 million hectares; China, 137 million hectares. China overtook the United States in agricultural production during the period 1979–81. China now far outpaces U.S. production of labor-intensive agricultural products, such as fruits and vegetables. Its output of meats is also much higher than that of the United States, and its production of land-intensive products is also slightly higher. But although China is both the world's largest agricultural producer and its largest agricultural consumer, the United States is the largest exporter of agricultural goods.

China has also overtaken the United States in terms of output of major manufactured goods (table 2-3). China ranks first in the world in its output of 170 manufactured goods but lacks world-famous brands.[68]

Table 2-3. *Major Manufactured Goods, China, Percentage and Rank, 2000s*

Unit as indicated

Manufactured goods and year	Share of world total (percent)	Rank
Chemical fiber (2005)	44.7	1
Iron ore (2005)	16.6	3
Energy[a] (2006)	14.5	1
Coal (2006)	39.4	1
Hydropower (2006)	13.7	1
Petroleum (2006)	4.7	5
Electric power[b] (2006)	15.1	2
Gold ore (2005)	9.1	4
Silver ore (2005)	12.4	3
Lead ore (2005)	32.7	1
Zinc ore (2005)	22.2	1
Tin ore (2005)	41.3	1
Tungsten ore (2005)	87.0	1
Molybdenite (2005)	19.0	3
Vanoxite (2005)	29.2	2
Manganese ore (2005)	10.5	4
Color TVs (2005)	40.5	1
DVDs (2005)	79.1	1
Digital cameras (2005)	52.9	1
Mobile phones (2005)	39.7	1
Computers (2005)	84.0	1
Iron (2005)	35.5	1
Raw steel (2005)	30.9	1
Copper (copper concentrate) (2004)	13.7	2
Aluminum (2004)	22.1	1
Lead (lead concentrate) (2004)	26.1	1
Zinc (2004)	25.2	1
Beer (2005)	19.1	1
Sulfuric acid (2005)	23.6	1
Urea (2002–03)	26.8	1
Phosphate (2002–03)	23.3	2
Granulated sugar (2005)	7.7	3
Paper (2005)	15.3	2
Synthetic rubber (2005)	13.5	2
Metal working machine tools (2006)	11.8	3
Motor vehicles[b] (2005)	8.6	4
Shipbuilding contracts (2006)	27.2	2
Ships built (2006)	15.0	3

Source: Yano Tsuneta Kinenkai, *Nihon kokusei 2007/08* (Tokyo: Nihonhyoronsha); World Bank, *World Development Indicators 2009;* British Petroleum, "BP World Energy Statistics," June 2007.

a. United States is 14 percent.

b. United States is largest exporter.

MEASURE 2, MARKET EXCHANGE

One can also use the market-exchange-rate method of measuring the GDP gap between China and the United States. This method, however, may exaggerate the difference between the two countries' respective GDPs, especially with regard to the service industry. Because living costs and production costs in the two countries are often quite different, simply using the market exchange rate to compare their GDPs will likely fail to account for people's real welfare. The market-exchange-rate method of measuring GDP leads to serious GDP uncertainty. The statistics below are from the World Bank, which uses the 2000 U.S. dollar as the standard unit of measurement. The three major components of GDP are examined individually before a holistic account is given of the GDP gap between the two countries (table 2-4).

In terms of the added value of agricultural production, China overtook the United States in 1980, after which (by 2005) its percentage of the world total rose to 16.97 percent. I estimate that this percentage could reach 23.0 percent by 2010, about 2.8 times greater than the United States percentage (8.1 percent). The biggest gap between the two countries exists in labor productivity, which will take a long time to narrow.

In terms of value-added industrial production, China's output constituted less than 2 percent of the world total in the 1980s, failing to rise until the early 1990s. By 2000 it comprised 6.30 percent of the world total, rising to 9.00 percent in 2005 and 12.09 percent in 2008. As of 2008 China ranked second in the world, trailing only the United States (21.14 percent). Between 1980 and 2008 this gap between the two countries was narrowed from a factor of 10.33 to a factor of 1.75.

In terms of the added value of the service industry, the gap between the United States and China is even larger, though it is in the process of being narrowed. In 1980 value added in the U.S. service industry was sixty-five times that of China. This gap was narrowed to a factor of 7.52 by 2008. It is important to point out that the exchange-rate method of calculating may both exaggerate the output of the United States and underestimate the output of China. The price of labor-intensive services in the United States is generally much higher than in China, and the output of the U.S. service industry may be exaggerated. Meanwhile, China's own National Bureau of Statistics seriously underestimated the actual output of China's service industry. Its first national economic survey in 2004 determined that the actual GDP value was 2.3 trillion yuan more than the original estimate. The added value of the service industry (service trade) had been underestimated by an amount equal to

Table 2-4. *GDP and the Added Values of Three Economic Sectors, China and the United States, Selected Years, 1980–2008*[a]

Unit as indicated

	1980	1990	2000	2005	2008
China GDP (percent of world total)	1.04	1.85	3.77	5.20	6.64
U.S. GDP (percent of world total)	29.06	29.40	30.75	30.39	28.79
U.S./China ratio	28.03/1.00	15.87/1.00	8.15/1.00	5.85/1.00	4.33/1.00
China agricultural output (percent of world total)	9.45	12.97	15.54	16.97	18.27
U.S. agricultural output (percent of world total)	6.93	7.58	9.80	8.70	8.97
U.S./China ratio	0.73/1.00	0.58/1.00	0.63/1.00	0.51/1.00	0.49/1.00
China industrial output (percent of world total)	1.16	2.22	6.30	9.00	12.09
U.S. industrial output (percent of world total)	24.24	23.21	25.16	23.88	21.14
U.S./China ratio	10.33/1.00	5.41/1.00	1.91/1.00	1.33/1.00	1.75/1.00
China service industry output (percent of world total)	0.53	1.23	2.39	3.02	4.39
U.S. service industry output (percent of world total)	34.89	34.06	34.50	33.91	32.98
U.S./China ratio	65.29/1.00	27.65/1.00	14.43/1.00	11.22/1.00	7.52/1.00

Source: World Bank, *World Development Indicators 2010*.
a. Calculated according to the constant price of the dollar in 2000.

2.13 trillion yuan, accounting for 93 percent of the adjustment. Bureau chief Li Deshui explains that the main reason for the modification was that many statistics from the service industry were not covered in the original report.[69] Although China's labor productivity in the service industry is much lower than that of the United States, the number of individuals employed is more than twice the U.S. figure and continues to grow rapidly.[70]

In terms of aggregate GDP, the gap between the two countries was as high as a factor of 28.0 in 1980; the factor was 8.1 by 2000 and 4.3 by 2008. The most glaring difference between the countries' GDPs is in the service industry. In 2009 the American service industry accounted for 76.9 percent of its GDP, while the Chinese service industry made up only 42.6 percent of China's GDP.[71] Thus China has to accelerate the development of its service industry, especially its modern services, if it is to further narrow its GDP gap with the United States.

How can China narrow its GDP gap with the United States given a constant yuan-to-dollar exchange rate? In 2005 America's GDP was approximately 5.5 times that of China, meaning that it is likely to take China a long time to catch up to and surpass the United States in economic might. If U.S. GDP expands at a rate of 3 percent a year and China's expands at a rate of 8–9 percent a year, then by 2020 America's GDP will be between 2.35 and 2.70 times that of China and by 2035, 1.06–1.32 times that of China.

If one takes into consideration the appreciation of the yuan against the dollar, the future actual price of the yuan should begin to converge with the international price (PPP dollar price). Given a 3–5 percent appreciation of the yuan against the dollar, by 2015 the exchange rate will stand at 5.10–6.18 yuan to the dollar, and U.S. GDP will only be 1.99–2.60 times China's. If the yuan continues to appreciate at that same rate, by 2025 the exchange rate will be 3.13–4.60 yuan to the dollar. If the yuan appreciates at an annual rate of 5 percent, the earliest year that China could overtake the United States would be 2023, with 2024–26 being the most likely period in which China's economy would become the world's largest. If the yuan appreciates at a rate of 3 percent annually, China could overtake the United States as early as 2027, or more likely between 2030 and 2035. Ultimately, the year China will replace the United States as the world's leading economic power depends not only on China's GDP growth but also on U.S. GDP growth and on the appreciation rate of the yuan.

Measure 3, the PPP Method

According to Angus Maddison, PPP, not exchange rate, is the GDP calculation method best suited to make cross-country comparisons. This is in part because one cannot compare nontradable goods or services using the exchange-rate method. Maddison also believes that the exchange-rate method seriously underestimates China's GDP. For example, according to the exchange-rate method, China's GDP was only 15 percent of U.S. GDP in 2003. The PPP method, however, suggests that China's GDP was 75 percent of the U.S. figure.[72]

Two variables affect the final results of these calculations. The first is the initial calculation of the GDP gap. There is some disagreement, for instance, on whether U.S. GDP is 1.2 or 1.4 times that of China. The other factor is China's future growth rate, estimates of which fall between 8 and 9 percent. The most likely scenario has China catching up to the United States sometime between 2012 and 2015. Many foreign scholars studying China's economy reached this conclusion, although they use different methods in their calculations.[73]

As early as March 1955 Mao was thinking about the prospect of the Chinese economy catching up with or overtaking that of the United States.[74] At the National Congress of the Communist Party of China, he said that for a country in the Orient with a population of 600 million to make socialist revolution, to change its face and the course of its history, to accomplish its basic industrialization and the socialist transformation of agriculture, handicrafts, capitalist industry, and commerce in a period of roughly three five-year plans, and to catch up with or surpass the most powerful capitalist countries in the world in several decades, it would inevitably encounter difficulties as great as or perhaps even greater than those encountered in the period of the democratic revolution. He later told participants at the conference that by "the most powerful capitalist countries" he meant the United States. In August 1956, at the Eighth National Congress, he reiterated this line of reasoning.[75] America's GDP in 1955 was 5.15 times that of China.[76] Although Mao Zedong used steel output as the indicator of economic power, he meant that the United States was the nation for China to catch up with in the twenty-first century. It now seems that Mao Zedong's grand strategy for China is on the verge of being realized.

China overtaking the United States in terms of GDP, regardless of how it is calculated, is inevitable. Indeed, it is simply a matter of time. Yet even when China can boast the largest economy in the world, its per capita GDP will remain far behind that of the United States. Thus, as Mao said in 1949, China will have taken only the first step in its "long march" to catch up with the United States. Future generations of Chinese leaders will determine what China must do to accelerate the pace of catching up and overtaking the United States in per capita GDP.

If China sustains its current level of growth (8–9 percent) through 2020, it will have, over the course of 42 years, expanded its GDP by a multiple of 41.5–46.4. This would be equal to total GDP expansion (45.8 times) of twelve Western European countries over the course of 180 years (1820–2000).[77] This is an economic miracle, the key to which lies in the consistency and sustainability of Chinese economic growth.[78] Of course, the key to such consistency and sustainability lies in the realization of a "great order under heaven" in China, meaning the existence of social and political stability and the continuity of the reform and opening policy.

As long as these "great order under heaven" conditions remain, China will continue to build upon its economic miracle. The United States maintained an annual per capita GDP growth rate of a modest 1.7 percent in the 178 years from 1820 to 1998. Over that period, per capita income rose from around $1,200 to $30,000 (in 1990 dollars). This process made America the world's largest economy.[79] Is there any doubt that China can do the same? I believe that China's future generations will be wiser and more resourceful than we are.

CHAPTER THREE

DEMOGRAPHIC CHALLENGES: AN AGING SOCIETY AND RAPID URBANIZATION

I n addition to being the most populous country in the world, China is an underdeveloped country that began the process of industrialization late, started from a low level of economic development, and features a transitioning population. When the People's Republic was founded in 1949 its population statistics were typical of countries at its level of development: high birthrates, high mortality rates, and low growth rates.

According to a survey conducted by Chinese scholars, the country's mortality rate before 1949 was roughly 25.0 to 33.0 per thousand, peaking at 40.0 per thousand.[1] At the same time, the birthrate was as high as 25.0 to 38.0 per thousand, but this number is an overstatement, because 20.0 to 25.0 per thousand of newborns died as infants. The UN population data bank shows that China's birthrate was 43.8 per thousand in 1950–55, with a mortality rate of 25.1 per thousand.[2] Both of these numbers were higher than the world average of 37.2 per thousand and 19.5 per thousand, respectively. Total fertility rate (TFR) of 6.11, natural population growth rate of 1.87 percent, and infant mortality rate of 195 per 1,000 all far outstripped world averages.[3] The average life expectancy, however, was only 40.8 years, 5.8 years less than the world average. In general, China's demographic conditions were typical of other traditional agrarian societies.

Since the establishment of the People's Republic of China, a series of fundamental changes have taken place in Chinese society. The first, marking the end of the prolonged civil war that split the country for several decades, was the replacement of chaos with order and the inauguration of a period of peaceful construction. The second was the commencement of economic development and the steady improvement of people's lives, culminating with the attainment of basic necessities for both urban and rural residents. As the economy improved, advances in medicine and improved health conditions effectively controlled the spread of many serious and communicable diseases. Then with the implementation of the family planning policy in the 1970s and broad knowledge of and access to birth control and contraception came a rapid transition in the structure of the modern low-income population. A significant drop in the death rate (and particularly in the infant mortality rate) resulted in a sharp increase in the mean life expectancy. After the establishment of the family planning policy, the birthrate and fertility rate began to drop sharply, allowing China to complete its demographic transition much faster than most industrialized countries.

According to UN population data, China's birthrate in the period 2000–05 was 14.0 per 1,000, its death rate 6.6 per 1,000, and its natural growth rate 0.7 percent. All of these figures were lower than the world averages of 21.2, 8.6, and 1.3, respectively. In addition, the total fertility rate was 1.8, much lower than the world average (2.8), suggesting that China is transitioning to a demographic structure the Japanese call *shoshika,* or "a society without children." The infant mortality rate was 25.6 per thousand, half of the world's average of 51.7 per thousand. The mean life expectancy was 72.0 years, longer than the world average of 66.4 years. All of these statistics indicate that China's population has entered a period in which its reproduction characteristics are typical of a modern society. Apart from the infant mortality rate and the mean life expectancy at birth, most of these demographic indicators were identical or very close to those of developed countries. Furthermore, it is expected that by 2020 China's infant mortality rate and mean life expectancy at birth will reach levels commensurate with developed countries, although per capita income will still lag far behind.

In the 1980s China's family planning policy produced a demographic dividend (*renkou hongli*), in which faster economic growth is attributable to the population having a higher percentage of working-age people.[4] This demographic factor has been favorable to China's sustained high-speed economic growth. As it moves into the twenty-first century, China must confront two major demographic challenges (in addition to improving its per capita income): a declining population of children and a burgeoning elderly population. Unlike the world's developed countries, which typically "grew rich before growing old," China is growing old before it grows wealthy. The demographic dividend that has served it so well will begin to fade around 2015, a shift that will undoubtedly have an impact on the country's economic growth. Meanwhile, for decades China has been in the throes of urbanization on an unprecedented scale, a process that will continue to impact virtually every aspect of Chinese society.

The PRC's Rapid Demographic Transition

Though China's demographic transition was completed in a timely manner, the process was not without rough patches. Indeed, the country experienced a special type of growth trajectory. Although there was a baby boom in the 1950s, a major famine in the years 1959–61 led to a baby bust. The second baby boom came on the heels of the famine of the 1960s. All of these affected China's population structure later on.

MODERN DEMOGRAPHIC TRANSITION, 1949–78

China's modern demographic transition began in the 1950s, when rapid industrialization led to rapid population growth. The population of the PRC stood at 542 million in 1949. It is expected to peak at around 2030, at which point an estimated 1.462 billion people will live in China. In eighty years the population will have increased by a factor of 2.7. Such population growth, both in terms of the absolute number of people and the multiplication factor, would far exceed population growth during any other period in Chinese history.

International comparisons show that China's total population in 1949 was much larger than those of developed countries when they launched their own periods of industrialization. China's 1949 population was 4.8 times that of the twelve Western European countries in 1820 (114 million), 52.0 times that of the United States in 1820 (10 million), and 15.7 times that of Japan in 1870 (34 million). At the beginning of China's demographic transition, the pressure posed by its massive population and the low education levels of its people were the largest obstacles to successful industrialization.[5] At the time, however, Mao Zedong was unaware of the country's demographic structure and did not expect the population to grow so quickly. He bristled at U.S. Secretary of State Dean Acheson's belief that "the Communist Party of China will not be able to solve its economic problems, that China will remain in perpetual chaos and that her only way out is to live on U.S. flour."[6] In contrast, Mao said: "Of all things in the world, people are the most precious. It is a very good thing that China has a big population. Even if China's population multiplies many times, she is fully capable of finding a solution; the solution is production."[7]

During China's first baby boom, 1949–59, the average birthrate was close to 30 per thousand and the cumulative number of births reached 230 million. Simultaneously, the mortality rate dropped sharply from 18 per thousand to less than 10 per thousand, the natural growth rate exceeded 20 per thousand, and the aggregate population grew by 123 million people.

During this period when the population was large and growing quickly, both the employment figures and rates of grain production increased; hectares of arable land peaked in 1957. Thereafter, these figures declined dramatically, becoming a significant hindrance to China's industrialization, urbanization, and modernization. It was impossible for Mao Zedong to overcome these challenges. As a result, he was forced to make some modifications to his "approaches to population," which were often self-contradictory and constantly changing.[8] Although the Chinese Communist Party (CCP) leaders Deng Xiaoping and Zhou Enlai favored birth control, people in positions of political influence outside of the party, including Ma Yinchu, Sun Benwen, and Fei Xiaotong, proposed family planning policies, none of which were endorsed or adopted by the state.

In the wake of the disastrous Great Leap Forward, China's birthrate dropped into the negative numbers. Once the economy began to recover, however, a second baby boom began. In the decade 1962–72 an average of 26.66 million people were born each year, with the cumulative number of new births reaching 270 million and the total population growing by 194 million people. At the same time, the first baby boom cohort (1950s) entered the workforce, making employment scarce in both urban and rural areas. In 1973 Mao proposed the establishment of a family planning office to organize family planning work across the country. In February 1974 he admitted for the first time that "the population of China is much too large." That same year he wrote in a note on a government document, "it is imperative to control the population" and "mankind must control itself so as to grow in a planned manner."[9] Thereafter, the Chinese government began to push the family planning policy throughout the country.

China's total fertility rate then decreased from 4.98 in 1972 to 2.75 in 1979, a decrease of 45 percent in a span of seven years. The momentum of excessive population growth was slowed, and the population structure began to change, with a shift to a lower birthrate, death rate, and natural growth rate. These changes to a large extent provided a demographic dividend, which benefited the country's subsequent economic takeoff.

Demographic Dividend, 1978–2020

In March 1979 Deng Xiaoping made it clear that the Chinese population problem was of strategic importance. He categorized the basic national situation as "large population, poor foundation, and limited arable land."[10] A year later, the CCP Central Committee made family planning a basic national policy and officially released the "one couple, one child" directive.[11] In 1982, in his report to the Twelfth National Party Congress, CCP General Secretary Hu Yaobang stated, "In China's economic and social development, population issues will always remain an extremely important problem. Family planning is a basic national policy of our country."[12]

Although the total fertility rate was low in the 1980s, it was still higher than the replacement level (2.1). Beginning in 1986 China experienced its third baby boom, although the birthrate was low compared to previous booms. The average number of newborns in the period 1986–90 was 24.19 million, meaning a population growth of 16.81 million a year. During this period, Zhao Ziyang, who was at first premier of the State Council and then general secretary of the CCP, advocated for allowing rural women to give birth to a second child if the first was female. This was considered a "small opening" in the one-child policy, instituted because of farmers' reluctance to plan their families. The decision had an immediate effect on the total fertility rate, pushing it up from 2.20 in 1985 to 2.42 in 1986 and 2.59 by 1987. This number then leveled off at around 3.0 in rural areas.[13] The UN population databank shows that China's total fertility rate was 2.63 during the 1985 to 1990 period, higher than the replacement level.[14]

After Jiang Zemin became CCP general secretary, he began to review China's population situation and returned to the basic national policy set at the Twelfth National Party Congress. At the First Family Planning Forum in 1991, he observed, "Family planning concerns the destiny of the country and if we do not strictly control the population growth and do not take family planning as a long-term major strategic policy decision, it will be impossible to improve the quality of our population." On May 12, 1991, the CCP Central Committee and the State Council issued the document "Decisions on Strengthening Family Planning and Strictly Controlling Population Growth," which made it clear for the first time that the "top leaders of the Party and government at all levels must be held responsible" for carrying out the task of family planning. Starting in 1991 the Central Committee and the State Council called a forum of top regional party and government leaders (provincial party secretaries and governors) to discuss family planning at the end of the annual National People's Congress and Chinese People's Political Consultative Conference. Later the forum was expanded to cover the topics of population, resources, and the environment.

Since the 1990s China has experienced a total fertility rate lower than the replacement level. It was the first of the populous developing countries to embrace a modern population structure characterized by a low birthrate, low mortality rate, and low natural growth rate. The UN population databank shows that China's total fertility rate dropped to 2.10 between 1990 and 1995, lower than the replacement level. Although the government maintains the family planning policy set in place thirty years ago, today's extremely low fertility rate is due in large part to economic development, as rising per capita income level, the acceleration of urbanization, increasing educational levels, and more flexibility and competition in the labor market have heightened the opportunity cost of having children. As a result, China's total fertility rate has dropped below that of even the United States and other developed countries.[15]

There are four defining characteristics of China's demographic transition. The first is that the transition took place when per capita income levels were still quite low. The second is that this transition took place very rapidly—in about fifty years rather than the nearly a hundred years that developed countries took to complete their transformations. China's transition was also much faster than those of other developing countries with similar income levels. Its third defining characteristic is that the transition was accompanied by coercive social change instead of the natural process of economic development. In other words, the transition was pushed forward by the government's implementation of the nationwide family planning policy, excluding areas inhabited by minority people. The fourth is that this rapid demographic change enabled the early arrival of a short-lived demographic dividend period, which in turn presages the upcoming demographic debt period.

The demographic dividend period delineates the time span during which China's working-age population is equal to or greater than 60 percent of the total population. It can be graphed as an inverted U curve, divided into three stages. In the first stage the demographic window opens and the working-age population rises steadily. The peak of the curve is reached when the working-age population as a proportion of the total is at its highest. The demographic window closes when the working-age population as a percentage of the total number of people begins to drop steadily. The exact shape of the curve varies from country to country, as they all experience demographic changes in different ways and at different rates. The window of demographic dividend can stay open for up to forty years, depending on the rate of fertility decline, at which point aging closes it. For example, Japan's window was 1955–95; Italy's, 1975–95; India's will be 1975–2035. Other poor countries in Sub-Saharan Africa, South Asia, the Middle East, and North Africa are just about to see their windows of opportunity open.[16]

China's demographic window opened around 1980, when the working-age population accounted for 59.8 percent of the total population. At that time, the number of individuals in the working-age population numbered 586 million, 3.86 times that of the United States (152 million), and was on the rise. China entered the demographic dividend period—in terms of population size, quality, and structure—earlier than India. Indeed, India lagged twenty years behind, not entering the demographic dividend period until after 2000, when the working-age population came to account for 60.7 percent of the total. Although an abundance of labor enabled sustainable development, continued GDP growth, and increases in private savings in China, it also led to tremendous pressure on the government to ensure high levels of employment. The government has worked hard to address this issue.

China will reach the peak of the dividend period in the years 2015–20, when the working-age population will account for 71.5–71.9 percent of the total. Others give different estimations: Wang Feng and Andrew Mason predict that China will exhaust its dividend in 2013.[17] By 2020 China's working-age population will be 996 million people strong, 4.45 times the size of the U.S. workforce (212 million). The demographic window will begin to close around 2020, when China's working-age population will begin to decline. It will reach 61.4 percent by 2050, the same proportion as that of the United States. In sum, China's demographic dividend period will last for about seventy years, from 1980 to 2050. Like China, the U.S. demographic dividend period will come to an end around 2050. Unlike China's, however, the U.S. demographic window opened in 1950, giving that country a full hundred years before the end of the dividend era.

An Aging Society: Fewer Children and More Seniors

Sustainable population development occurs when the TFR is kept roughly at the replacement level (about 2.0). It is not the case that the lower the TFR, the better. A total fertility rate that is too low (less than 1.8) or too high (greater than 2.1) makes population development unsustainable. A TFR lower than 1.8 is classified as low fertility and leads to a decline in the child population. A TFR of less than 1.5 is classified as very low fertility and leads to a serious decline in the child population. A TFR of less than 1.35 is classified as the lowest fertility and leads to an even sharper decline in the child population.

China's TFR has been declining steadily since the beginning of the 1990s. Indeed, it has dropped below the replacement level (2.1) and entered a period of low fertility, in which the size of the child population is declining. According to the UN population databank, China's TFR was 1.8 during the period 1995–2000, leading to a consequent drop in the population of children.[18] The TFR dropped further, to 1.5 in the years 2000–10, lower than that of the United States (2.10). The TFR in Chinese cities and towns was lower than 1.3, leading to a sharp decline in the child population. In rural areas, meanwhile, it was just below 1.8, not as severe but still enough to prompt the stage of low fertility that will lead to a decline in the child population. The cohort of only children born after 1980 has entered the marriage stage. According to a 1 percent population sample survey, there were 155.89 million only children aged 0–30, accounting for 29.30 percent of the population of similar age.[19]

China's total number of children has dropped sharply since the 1980s.[20] The population of children stood at 341.56 million in 1982. Twenty-three years later, in 2005, it had dropped to 265.43 million, a decline of 76.13 million, or 22.3 percent. By 2009 it had dropped to 225.17 million, a decline of 116.39 million, or 34.1 percent, since 1982. As a result, the number of people aged fifteen to thirty will be reduced by 76 million during the period 1997–2020. The percentage of the world's children living in China has also declined.[21] In 1980, 22.4 percent of all children lived in China. By 1990 that share had dropped to 18.4 percent, subsequently falling to 17.2 percent in 2000 and 15.7 percent in 2004. The number of students enrolled in primary schools has also dropped steadily.[22] In 1997 the primary school population was 139.95 million. By 2009 it was 100.72 million, a drop of 39.23 million or 28.0 percent.

Demographically speaking, a country can be described as an aging society when people aged sixty-five or above make up more than 7 percent of the total population (as in China). When the elderly make up more than 14 percent of the total population, the country is termed a seriously aging society (as in Japan in 1995). When this age group makes up more than 20 percent of the population, the country has a super-aging society (as will be true of Japan in 2015). In 2000 Chinese citizens aged sixty-five or above made up 6.96 percent of the total population, making China an aging society. In achieving this distinction, China became the first developing country to become aged. As of 2009, 8.5 percent of individuals in China were sixty-five or older.[23] An elderly population that stood at 48.81 million in 1982 had increased to

113.45 million by 2009. Its average annual growth rate of 3.2 percent was much higher than the growth rate of the total population (1.0 percent).[24] Retired people in cities and towns increased from 23.01 million in 1990 to 50.88 million by 2005, averaging an annual growth rate of 5.4 percent, much higher than the growth of the total urban population (4.2 percent).

The UN population data bank shows that by 2020 China's population aged sixty and above could reach 239 million, or about 16.7 percent of the total population, twice as many as lived in the country in 2008.[25] By that time, the number of individuals aged sixty-five or above will have increased to 167 million, or 11.7 percent of the total population, with the number of people aged eighty or above increasing to 28.15 million, or 2.0 percent of the total population. By 2050 people aged sixty-five or above will number 330 million, or 23.3 percent of the total population, and people aged eighty or above will increase to 101 million, or 7.2 percent of the total population.

The rapid growth of the aged population has resulted in a steady rise in the dependency coefficient, or ratio of laborers to elderly persons. In 2000 an average of 9.1 laborers supported one elderly person. By 2020 there will be 5.9 working-age people for every elderly individual. By 2050, 2.7 working-age people will support one elderly individual. These statistics suggest the coming of a demographic debt period in the PRC. China would be wise to draw lessons from Japan. The TFR of Japan dropped below the replacement level (2.1) in 1973, falling to 1.29 by 2004. During this time, the percentage of the population over age sixty-five rose from 7.6 to 20.0 percent. As of 2004 Japan's absolute population was dropping. It is expected that the total population of Japan will drop to 100 million by 2050, with people aged sixty-five and above making up 35.7 percent of the total.[26] An aging society with fewer children has become one of the principal reasons that the Japanese economy features inadequate levels of internal demand and lacks the strength to grow.

China has contracted the so-called Japanese disease: a rapid decline in the child population accompanied by the rapid rise of the aged population. China, however, caught the malady twenty years later than Japan, even thirty years later in terms of the working population aged fifteen to forty.[27] The biggest difference between the two countries in this respect is that Japan attained wealth before the child population began to decline and the aged population began to rise, while China will likely grow old before getting rich. This is an issue of great concern to Chinese scholars and policymakers alike, and it is likely to fundamentally change China's economic development trajectory from high growth levels to low growth levels. The fifty years between 2000 and 2050 will be a crucial period for China as it looks to become a truly prosperous and harmonious modern country while coping with the challenges posed by an aging population.

Adjusting Population Control Policies

The size of its population has always been an issue for China. Overpopulation is both a development problem and a very complicated social problem. For this reason, population planning has become a strategic part of development. The slow nature of population planning makes it impossible for China to fix this problem by adopting a short-term policy.

In retrospect, if we calculate from the beginning of the 1980s, solving the population problem will take at least two generations (or fifty years). Given the characteristics of population planning and the logical effects of adjusting policy, the solution will probably require two steps. The first step is the implementation of the family planning policy of 1980–2005 mandating one couple, one child in major cities and developed areas in order to control population growth and adjust the population structure. The second step covers 2005–30, when the state should allow one couple, two children, all the while focusing on controlling the size of the population through the adjustment of the age structure as well as the rural and urban population structures. In 1980 the Central Committee stated that after thirty years the problem of population growth would be eased, allowing for the adoption of different population policies.[28] The time has come for the implementation of such policies.

Many among China's only-child generation have entered the childbearing age. According to the current family planning policy, all couples both of whom were only children may have two children. I advocate further relaxing this policy, allowing a couple in which only one of the two was an only child to have two children and allowing a remarried couple in which both individuals are only children or one is an only child to have two children. In other words, I agree with Peking University professor Zeng Yi's proposal: "One couple, two births, and late marriage for a soft landing."[29] This policy would adjust the future population structure and the male-to-female ratio, lowering the future dependency coefficient and easing the problems posed by pension shortages and the loss of China's labor power advantage.

To cope with the problem of a smaller working-age population, it is imperative both to encourage self-employment and to expand the supply of labor, especially of high-quality human resources. On the other hand, it is also important to reduce the number of retired people so as to ease the pressure on China's social security program. To accomplish this, China must extend the retirement age for men and women. Since the 1950s the legal retirement age has been fifty-five for women and sixty for men. The rise in the mean life expectancy at birth, however, has made it necessary to modify this outdated retirement policy so as to cope with the rapidly aging population.

In extending the retirement age, three steps should be taken. The first step should apply to civil servants, professionals, corporate managers, and management personnel, whose retirement age should be extended to sixty-two for both men and women. The retirement age for other women should be extended to sixty. The second step is to extend the retirement age to sixty-five for all working people. The retirement age for those in private enterprises should be flexible. There should be no regulations as regards a retirement age for these employees. Instead, they should be encouraged to continue working. The third step will affect part of the service industry, in which a flexible policy should be introduced.

It is necessary to raise the minimum age at which workers can begin receiving pensions and implement a more flexible employment system, including self-employment, nonregular employment, contracted employment, and family employment. In these fields there will be no retirement age regulations. In the long term, the most fundamental way to deal with the challenges of an aging population is to continue intensifying investment in human capital.

Fundamental changes have taken place in China's national conditions, necessitating changes in its population planning strategy. The national population policy should be based on the dynamic changes and characteristics of China's national conditions. It must also be based on a long-term, forward-looking strategy so as to forestall the creation of new burdens and difficulties to be dealt with by future generations.

Urbanization: The Main Driving Force of Future Development

Cities are the areas most in need of greater production and modernization. Urbanization has already contributed significantly to China's economic growth, and it will remain an important source of economic growth in the future. In a global and historical context, China is a latecomer to industrialization and urbanization but has largely caught up in both categories and is enjoying the advantages of being a latecomer to the process. Unlike many developing countries, whose major cities feature slums, China should build, and has the capacity to build, people-centered and harmonious socialist cities.[30]

China's urbanization process is unique. Its large base population ensures that the impact of China's urbanization will be felt by the world. Indeed,

China's rate of urbanization currently ranks first, making it not only an engine of world economic growth but also an engine of worldwide urbanization. In addition, urbanization only picked up speed after reform and thus bears obvious catching-up features.

THE PLANNED ECONOMY PERIOD: URBANIZATION REPLETE WITH TWISTS AND TURNS

The process of Chinese urbanization has been long and slow. The urbanization rate was 3.0–3.8 percent between 1000 and 1820, rising to 4.4 percent by 1890. This was dramatically slower than in Europe, where urbanization rates were 12 percent in 1820 and rose to over 30 percent by 1890. In Japan the percentage of the total population living in cities stood at 16 percent in 1890. Between 1900 and 1938, Chinese people were free to move, and the growth rate of the urban population was twice that of total population growth.[31] As of 1938 China's total urban population in cities with a population of over 50,000 reached 27.30 million, accounting for 5–6 percent of the national total population (500 million).

The rate of urbanization was very low when the PRC was founded. Of the 542 million people in the country, 484 million, or 89.4 percent, lived in rural areas. The urban population accounted for only 10.6 percent. At the time there were only 132 cities in China. If calculated according to the nonagricultural population in cities, the real percentage of the urban population was only 5.1 percent.[32] Though China's total population at the time constituted 21.7 percent of the world total, its urban population made up only 8.5 percent of the global aggregate. This was the starting point of China's modern urbanization process. Even during the planned economy period, the growth rate of China's urban population was higher than that of the United States and the world average. During this period, the annual growth rate of China's urban population was as high as 3.85 percent (table 3-1).

The 1950s was a period of fast growth. Over the course of the decade the urbanization rate rose by 8.5 percentage points, while the urban population had an average annual growth rate of 7.8 percent. By the end of the decade, the relative gap between China and the United States in urbanization had narrowed by 2.7 percentage points. All of this was the result of the rapid rehabilitation of industry during the period. The 1960s marked a period of decline, with the urbanization rate dropping by 2.3 percentage points and the gap with the United States widening to 5.9 percentage points. In the ten years from 1960 to 1970, due to China's economic structure and the tight controls presented by the household registration system, the pace of urbanization was

Table 3-1. *Urban Population Growth, China, the World, and the United States, Selected Periods, 1950–2030*

Percent

Period	China	World	United States
1950–80	3.85	2.91	1.75
1980–2000	4.47	2.50	1.41
2000–30	2.19	1.86	1.16
1960–70	0.99	2.94	1.73
1950–2030	3.38	1.44	1.44

Source: UN Department of Economic and Social Affairs, *World Urbanization Prospects: The 2008 Revision.*

slow, with the overall urbanization rate growing at an average annual rate of less than 1 percent.[33] This number was far lower than both the world average of 2.94 percent and the United States average of 1.73 percent. Chinese cities at the time were unable to accommodate many people, so much so that the government even sent educated youths to rural areas. This in fact restricted the further development of urbanization. Thus the speed of urbanization experienced considerable fluctuations.

From the early 1970s through to the beginning of reform and opening there was little progress in urbanization, with the rate stalling at 2.0 percentage points. The annual growth of the urban population increased to 2.87 percent from 1970 to 1980, a period during which global urbanization accelerated while China's crept forward at a snail's pace. Only 17.4 percent of Chinese people lived in cities in 1975, 20.5 percentage points lower than the world average of 37.9 percent.[34] These were the initial conditions for urbanization at the beginning of reform and opening.

Urbanization after Reform and Opening

China's urbanization was put into high gear after reform and opening began. From 1980 to 2000 the overall percentage of China's population living in cities rose by 16.8 percent, narrowing the gap between the PRC and the United States by 11.4 percentage points. During this period China's urban population grew at an average annual rate of 4.47 percent, twice as fast as the world average and three times faster than the United States (see table 3-1). China's urban population in 1980 constituted 11.0 percent of the world's total urban population, 1.12 percent more than the United States (9.78 percent; table 3-2). By 2000 China's share of the world total stood at 16.09 percent, far

Table 3-2. *Urbanization, China and the United States, 1950–2030*
Unit as indicated

Year	Urban population (million)			Urbanization rate (percent)			Share of world urban population (percent)	
	China	United States	Ratio	China	United States	Gap	China	United States
1950	61.7	101.2	0.61	11.2	64.2	53.0	8.37	13.74
1960	130.7	130.3	1.00	19.7	70.0	50.3	13.12	13.08
1970	144.2	154.6	0.93	17.4	73.6	56.2	10.83	11.61
1980	191.4	170.3	1.12	19.4	73.7	54.3	11.00	9.78
1990	302.0	192.8	1.57	26.4	75.3	48.9	13.28	8.48
2000	459.1	225.3	2.04	36.2	79.1	42.9	16.09	7.90
2010	637.0	259.0	2.46	47.0	82.3	35.3	18.23	7.41
2020	856.0	290.7	2.94	61.0	84.9	23.9	20.34	6.91
2030	1,043.0	318.5	3.27	71.0	87.0	16.0	21.00	6.41

Source: National Bureau of Statistics of China, *Statistical Data of the 60 Years of New China* (2010); UN Department of Economic and Social Affairs, *World Urbanization Prospects: The 2008 Revision;* author's calculations.

higher than that of the United States (7.90 percent). The rapid increase in the rate of China's urbanization has had a significant impact on the trend of overall global urbanization.

A comparison of urbanization rates in China and the United States between 1980 and 2000 shows that while the United States entered a period of slow growth, China was experiencing the fastest urbanization in its history. At the beginning of this period China's urban population already exceeded that of the United States in absolute numbers, ranking first in the world. By 2000 China's urban population was twice as large as the U.S. figure. Although there is still a large rural population in China, these rapid urbanization trends mean that China will become the biggest consumer market in the world. The rise of income and living standards of the people has thus become the necessary condition for China, as the country with the biggest urban population, to overtake the United States.

The main reason for the accelerated pace of urbanization is the large-scale movement of China's rural population into cities and towns, the biggest population flow in the history of the world. Zhang Honglin of Illinois State University and Song Shunfeng of the University of Nevada estimate that about

174 million rural people moved to cities between 1978 and 1999, accounting for 75 percent of total urban population growth.[35] Since the 1980s China's urban population has entered into a stage of low birthrate, low mortality rate, and low natural growth rate. The added population in cities and towns has come mainly from rural areas. This includes population migration and population movement. Population migration includes those directly recruited from rural areas, graduates of universities and intermediary technical schools, and demobilized soldiers and military personnel who quit active service. Population movement refers to people who have left their original places of household registration and moved to other areas for temporary residence or for other activities without changing their permanent residence. Most farmers working in cities belong to this category.

According to the Ministry of Agriculture, 206.75 million rural people moved to cities and towns between 1982 and 2000, amounting to 45 percent of the urban population and 84.6 percent of the newly added population in cities.[36] Over the same period the number of rural laborers moving to cities and towns was 109.6 million, amounting to 45.8 percent of the total labor force in cities and towns, or 94.3 percent of newly added labor. The movement of total population and labor power expanded swiftly over time: 8.14 million and 4.01 million, respectively, in the 1980s; 14.14 million and 7.75 million in the early 1990s; and 20.30 million and 11.46 million during the latter half of the 1990s.

Rapid urbanization during this period boosted demand for public services and social management, leading to major changes in China's economic and social structure and a weakening of the household registration system and other obstacles to urbanization. Despite this there was no fundamental change in the huge gap between urban and rural areas, a dichotomy that many refer to as one China, two societies. In addition, the structure of Chinese municipalities gradually changed from the urban-rural dual structure to a triple structure featuring cities, the countryside, and rural townships or urban nonregular employment institutions. Some even argue that a quadruple structure emerged, featuring cities, the countryside, rural townships, and migrant laborers in the nonregular sector of the economy.

Urbanization in the New Century: Opportunities and Challenges

After 2000, with a drop in the growth rate of the Chinese population, the growth rate of the urban population has also declined. Still, China's urbanization continues to accelerate. From 2000 to 2030 it is estimated that the average annual growth of China's urban population will drop to 2.77 percent. Even so, it will still be higher than the average urban growth rates in the United States and across the globe. This period will be of great importance for China in its attempt to leap forward as it urbanizes and builds an all-around well-off society. From 2000 to 2030 the percentage of Chinese citizens living in cities will rise by 34.8 percent, narrowing the gap between the PRC and the United States by 26.9 percent. This represents the fastest urbanization rate in history.

China's urban population as a percentage of the world total will reach its peak by 2020, when its urban population will make up 20.34 percent of the global aggregate (see table 3-2). It will remain first in the world, at 21 percent by 2030. The U.S. urban population as a percentage of the world's total will continue to drop, to 6.41 percent by 2030. This shows that although the process of urbanization in the United States is completed, it is just beginning to accelerate in the PRC. According to World Bank data, the world's average urbanization level was 47 percent in 2001, with that of lower middle-income countries being 46 percent and that of upper middle-income countries being 77 percent.[37] Even when the percentage of Chinese citizens living in cities reaches 71 percent in 2030, it will still be 27 percentage points lower than that of the United States, promising great potential for growth.

The urbanization process has pushed forward the construction of urban infrastructure. In 2007 total kilometers of roads in cities reached 246,100, far exceeding the planned goal of 180,000 kilometers, which was realized in 2002.[38] The road area on a per capita basis was 11.4 square meters, 10.0 square meters more than planned. The area in cities receiving central heating was 3.01 billion square meters, with 350.0 million square meters added, six times the planned amount. The natural gas supply was 30.86 billion square meters, with 6.38 billion square meters newly added. The annual supply of liquefied petroleum gas was 14.668 million tons, 2.031 million tons more than planned. The fuel gas penetration rate rose from 45.4 percent in 2000 to 87.4 percent in 2007, close to the planned goal of 90 percent. The areas of parks and greens per capita reached 9.0 square meters, an increase of 0.7 square meters and exceeding the planned goal of 8.0 square meters. Sewage water treatment capacity reached 103.365 million cubic meters, far exceeding the planned goal in the Eleventh Five-Year Plan. The process of accelerated urban infrastructure construction has also not been without problems. Among them are occupying too much land, coercive land removal and requisition, and a flood of projects designed by government officials to artificially enhance their performance and thus their image.

Throughout the next twenty years the acceleration of urbanization will effectively stimulate the movement of labor between town and country, bringing huge economic benefits. But in China the institutional arrangement of one country, two societies continues to restrict the free movement of labor. Such a system, designed to protect the minority, has not only harmed the interests of rural residents but also resulted in huge institutional opportunity costs, forcing the macroeconomy to pay a heavy price. The challenges faced by rural areas, farmers, and agriculture generally have become the most acute and obvious problems in China's contemporary economic development. The fundamental solution to these problems is to accelerate urbanization—that is, to release farmers from their rural status, invest in them, transfer them, reduce their overall population, and make them rich. In order to change the dual structure of citizenship in cities into a single, unitary structure, public services should be extended to all migrant laborers and their families. The key to achieving universal coverage in terms of education, health, housing, work compensation, and social security is to avoid the types of problems that have plagued other developing countries as they urbanize. These include the rise of urban slums, which are likely to become high-crime zones, the creation of major traffic bottlenecks, and the incessant widening of income gaps. China's urbanization should make cities more livable, more harmonious, and more sustainable in their development.

China's urbanization strategy has already shifted from encouraging the development of medium-size and small cities to encouraging the development of large and super-large cities. The Chinese government has set out to formulate a functional zone program.[39] The program is designed to bring about a strategic urbanization pattern by creating three super-large city groups (in the Bohai Rim area, the Yangtze River delta, and the Zhujiang River delta), each with a population of 60 million to 100 million, a number of extra-large cities with populations of 30 million to 50 million, and a number of large city groups of roughly 10 million to 20 million people each. These groups will be created around the foundation of current cities designated for major development along the two horizontal axes formed by roads and bridges parallel to the Yangtze River. Other cities and urbanized areas will serve as the main components on the three vertical axes along the coasts, the Beijing-Harbin and Beijing-Guangzhou railways, and the Baotou-Kunming railway. All super-large, extra-large, and large city groups will be distributed along the two horizontal and three vertical lines.

In terms of development goals, China's urban population stood at 622 million in 2009 and is expected to reach nearly 856 million by 2020, an increase of 37.62 percent. In terms of city space, urban areas will be extended from the current 79,300 square kilometers to 94,700 square kilometers, an increase of 19.4 percent. In terms of population density, the number of individuals per square kilometer will be increased by 20 percent, from the current 7,489 persons to 9,039 persons per square kilometer. In addition to increasing economic density (GDP generated per square kilometer of *land*, financial income, and the large growth of internal and external trade will intensify the agglomeration of quality factors such as number of scientific and technical personnel, specialized personnel, and university students per square kilometer as well as the amount of foreign direct investment and the number of R&D projects per square kilometer. Eventually, the major urbanized areas, which will occupy less than 10 percent of the total land territory, will hold about 60 percent of the population, generate over 80 percent of the economic aggregate, and account for over 90 percent of international trade. These new urban zones will not only drive future economic growth but also reshape China's economic geography.

Urbanization is a long historical process. It took more than a century for some Western countries to realize urbanization. Only Japan and the Republic of Korea realized urbanization in a relatively short period of time. It took Japan thirty-five years to raise the share of its population living in cities from 18 percent (1920) to 56 percent (1955). It took the Republic of Korea twenty years to raise its urbanization rate from 27 percent (1960) to 59 percent (1980).[40] As the most populous country in the world, China needs 10 million people to move to the city in order to raise its urbanization rate by a single percentage point. Therefore, it will take China a long time to realize urbanization. And yet, only once China truly realizes urbanization will modernization be within reach. In this sense, urbanization is the basic path to modernization.

The general strategy for urbanizing in China is to develop from a minimally urbanized country to a mid-level urbanized country, and from latecomer in terms of urbanization to a big power that ranks number one in terms of its urban population. The urbanization process in China has been complex, shifting from the promotion of urbanization to its restriction and back again to its encouragement. The type of urbanization being promoted has also changed, from an early focus on developing small cities and towns to a focus on developing extra-large city groups. By the end of the urbanization process, China's social structure will have expanded from a dual structure to a tripartite structure or even a four-part structure before transforming into a single, unitary structure. The ultimate goal will be the realization of social harmony and a well-off society.

CHAPTER FOUR

A Healthy China:
Progress and Problems

Health is the foundation of personal happiness, and a country's public health is often a major determinant of its rise or fall. The pursuit of a healthy life is a timeless universal value. Modern China was once labeled the sick man of East Asia, subject as it was to national subjugation and mass violence. After the founding of the PRC, however, the health level of Chinese citizens improved greatly, with the mean life expectancy of both men and women increasing dramatically. Still, China faces huge health challenges. The incidence of major destructive diseases has increased, and acquired immune deficiency syndrome (AIDS) is spreading. In 2003 a public health crisis broke out in the form of severe acute respiratory syndrome (SARS). One of the government's core objectives in building a well-off society over the course of the next twenty years is to ensure the general health of the people and enable a population of more than a billion to live healthier, longer, happier, and better lives.[1]

From the Sick Man of East Asia to a Giant in the East

Before the founding of the PRC, China was not only the poorest country in the world but also a laggard when it came to public health. Communicable diseases were widespread, mean life expectancy was short, and the mortality rate was high. Before 1949 the mean life expectancy at birth was only thirty-five years, roughly the 1820 level of Western Europe and lower than both the 1820 U.S. level and the 1950 world level (forty-nine years). The death rate was over thirty per thousand, the infant mortality rate was about 20 percent, and the maternal mortality rate was 1,500 per hundred thousand.[2] As a result of poor hygienic conditions, endemic diseases were extremely common. These health disasters posed a serious threat to the life and welfare of the Chinese people. Indeed, together they constituted the number-one cause of death in China. According to statistics, deaths from tuberculosis reached 250 per hundred thousand in cities during this period. Thanks to tremendous achievements in health and medicine since the founding of the PRC, however, China has erased its image as the sick man of East Asia.

The relationship between health and economic development in China may be divided into three stages. China's first health revolution took place in the period between 1949 and 1978. When the PRC was founded, tens of millions of people suffered from communicable diseases. There were 60,000 cases of smallpox in 1951; 1.04 million cases of measles in 1952; 2.93 million cases of cholera, 500,000 cases of leprosy, and 12 million cases of snail fever (schistosomiasis) in 1949. The number of people threatened by these diseases stood at more than a hundred million. The incidence rate of infectious diseases in the 1950s and 1960s was 3,000 per a hundred thousand people.[3] During this period, the absolute number of people suffering from infectious diseases ranged between 16 million and 22 million. With this in mind, the first health revolution sought to provide basic public health services, including prescription drugs and medical attention to help prevent, control, and eliminate communicable diseases. As a result of this campaign, public hygiene improved significantly, with many health indicators rising markedly, even past what one might expect for China's level of economic development. And because the people were not forced to pay high medical expenses, the general welfare of the Chinese people, especially those in poverty, improved dramatically. Over this period China's infant mortality dropped to thirty-five per thousand, and mean life expectancy rose to sixty-six years, outstripping developing countries with similar incomes. Real material quality of life surpassed per capita GDP. The noted Japanese economist Minoru Kobayashi remarked that although China's per capita income was not high, the Chinese people actually enjoyed a fairly high standard of living.[4]

During the second stage, spanning 1978 to 2003, mean life expectancy at birth increased from 67.8 years to 73.0 years. The period 1980–2000 witnessed relatively high growth in per capita income but relatively meager improvements to the Human Development Index (HDI). If calculated by constant price, per capita GDP grew at an average annual rate of 8.3 percent, twice as much as in the period 1950–80. The consumption level grew at an average annual rate of 7.1 percent, 2.3 times the 1950–80 rate. Mean life expectancy at birth, however, grew by an average annual rate 0.3 percent lower than the 1950–80 level. The infant mortality rate shrunk by an average annual rate of 1.6 percent, far lower than the 1950–80 rate. In sum, HDI grew at an average annual rate of 1.4 percent, also lower than the 1950–80 level. Mean life expectancy at birth rose to 71.4 years, reaching the average level of high-income countries (71.0 years). The infant mortality rate fell to 28.4 per thousand, and the death rate of children under five dropped to 37.0 per thousand in 2003, lower than the average of this age group in low-income countries (39 percent) but still higher than the average in middle- and high-income countries (26 and 35 per thousand, respectively). The HDI increased to 0.724, placing China in the lower-middle group.[5]

China's current major health development indicators are even better: higher than world averages and higher than those of most middle-income countries. According to the UN Development Program the worldwide average for 2002 mean life expectancy at birth was 66.9 years (it was 59.7 years for low-income countries, 69.7 years for middle-income countries, and 78.2 years for high-income countries).[6] China's was 71.4 years. By 2005 China's mean life expectancy at birth had climbed to 72.9 years. By 2007 its infant mortality rate had dropped to 15.3 per thousand. Moreover, the country's age-standardized death rate dropped to ten per thousand in the late 1990s, and the infectious disease incidence rate dropped to 267 per hundred thousand by 2006.[7]

By improving general medical conditions, China strives to increase life expectancy. This process, however, is difficult and slow. Comparing age-specific, disease-caused mortality rates between rural areas in 2003 and large cities in 2006, we find that although the death rates are nearly identical among people age fifty and younger, they are quite different among people age fifty and above, with much higher rates in rural areas than large cities.[8] If we view rural areas and large cities as occupying two different stages in China's health development, we can conclude that China has basically completed the transition from the first step to the second step in its health revolution. It is now incumbent upon rural areas to move from the second step to the third step, primarily by improving medical services and lowering the death rate of the older population.

The third stage began in 2003 and will last until at least 2020. With the outbreak of SARS in 2003, China began to shift from making economic growth the sole priority and also making health a priority. If the Chinese government is callous in its investment in and protection of health, the overall health of the Chinese people is likely to suffer. This may not be reflected in mean life expectancy, because mean life expectancy only shows the average duration of life and cannot be used to measure health. Rather, it will be reflected in the spread of chronic diseases. Such diseases may not result in a high death rate, but they can affect the ability of the people to work, thus adding to the country's social burdens. Needless to say, increased social burdens threaten the sustainability of economic development. The 2003 SARS outbreak sounded the alarm. It is therefore imperative to adjust the current strategy for long-term economic growth so that the economy and public health may develop in a mutually reinforcing manner.

A comparison of health indicators between China and the United States suggests that China's health revolution has yielded tremendous dividends for the Chinese public. From 1950 to 1955 China's mean life expectancy at birth was 28.1 years less than that of the United States, and its infant mortality rate was 167.2 per thousand higher. By 1970–75, however, China's mean life expectancy was only 8.3 years less than that of the United States, and its infant mortality rate was only 43.0 per thousand higher. These basic indicators show that the health of the Chinese people improved significantly during the Mao Zedong era. As of 2000–05, China's mean life expectancy at birth was only 6.3 years less than that of the United States, its infant mortality rate was reduced to 19.2 percent more than that of the United States, and the death rate for children under the age of five was 14 percent less than the U.S. figure. In other words, the gap in health indicators between China and the United States has steadily narrowed.

Public Health and Economic Development

Superior health outcomes can stimulate economic development. According to the health economics theory, improved health can extend the number of years a person can work and improve the workforce's ability to obtain and apply knowledge, thus increasing human capital and enhancing labor productivity.[9] Health can also affect economic growth by influencing the country's labor supply, educational levels, capital accumulation, demographic structure, and income distribution. In his study on the subject, the Columbia University professor Jeffery Sachs concludes that health was the biggest contributing factor to China's economic growth in the 1980s.[10] Without a doubt, increases in the health level of the Chinese people before 1978 provided a solid human capital foundation for its reform and opening economic takeoff. In 2002 the Nobel Prize–winning economist Amartya Sen used differences in health levels to explain the incongruent economic development stories of China and India. While India's health conditions had improved slightly, the betterment of China's health indicators in the years before reform and opening "created the social opportunities for the country to make dynamic application after the country marketized."[11]

According to my own calculations, these improvements to China's pre-reform health conditions contributed 0.92 percentage points from 1978 to 2000.[12] The direct contribution of education to economic improvement was 0.07 percentage points. If the two are combined into a human capital factor, they added about 0.99 percentage points to China's economic growth rate. Without this human capital as a foundation it would have been impossible for China to launch its economic takeoff.

In general, however, the improvement of public health conditions over the reform era has not been as impressive as China's economic strides. How is one to interpret the high rates of economic growth corresponding to minimal improvements to health conditions? In 1992 Amartya Sen, understanding the decline of the Chinese mortality following the implementation of economic reform, engaged in a debate with several British experts studying the Chinese economy.[13] Sen contends that economic growth does not necessarily bring about health improvements (as represented by declines in the death rate and increases in life expectancy). Rather, the distribution of economic resources was more important for improving health indicators. He says that the main reason there was a marked improvement in the health of the people during the Mao era was that food and medical resources (including rural medical services) were distributed equally. Correspondingly, one of the important reasons for the slower pace of health improvement after reform and opening is the increased inequality in social and economic development.[14]

When it comes to health, economic growth is a double-edged sword. It can lead to better material conditions for improving health standards, but it can also threaten or harm health by creating environmental pollution, population movements, and irrational consumption patterns.[15] This paradox is not unique to China; the same is true in the United States and Western Europe. When it comes to the relationship between the economy and health, the key is what development strategy is implemented and whether or not social and economic development policies improve health outcomes. Since reform and opening, China's economic development has been achieved at the expense of health. A single-minded pursuit of economic growth has been complemented by a neglect of the health of the people. This is the primary reason that economic development has been fast but improvements in health have been slow.

As regards the role of health in driving economic development, the foundation laid before reform and opening helped ensure a healthy supply of labor favorable for economic growth in the subsequent decades. Nevertheless, the contribution made by health to economic growth is analogous to a savings account. The big improvement to health care during the prereform era was equivalent to putting money into the account, but after reform began, slower improvements to health care have been like China drawing money out of the account instead of putting more money in. If such a situation continues, the account will fall into a deficit—a health deficit. It will be difficult to maintain health at a high level, especially if health risks grow.

China's Multidimensional Health Insecurity

Health insecurity refers to situations in which health risk factors are not effectively controlled or people are deprived of their basic health rights. Examples include a lack of access to basic health services, a lack of medical insurance, generally unhealthy living conditions or lifestyles, and a lack of basic health care knowledge. Health insecurity has become one of the biggest challenges to human security in China. With this in mind, health security has become one of the most important issues to address as China strives to build a healthy and well-off society.

My studies find that China has a large number of people subjected to various types of health insecurity.[16] Several key areas deserve serious attention.

Of the 5 billion cases of disease in cities and towns, nearly half had not received medical treatment by 2003. (A case in this instance is a measurement combining number of people and their work contribution.) According to a national health survey, cases rose from 4.3 billion in 1993 to 5.0 billion in 2003, growing by 16.5 percent, or 700 million, and averaging 3.63 times a year per case, an overall increase of 3.86 times.[17] Of the patients surveyed, 21.8 percent characterized their illness as not serious, 51.3 percent characterized their illness as mild to moderate, 24.1 percent characterized their illness as serious, and 2.8 percent replied that they could not tell.

In 2003 the cases of people who did not feel well but failed to seek treatment was 2.35 billion, accounting for 47 percent of the total. These cases neither went to the hospital nor asked for a doctor. The number of those who did not receive medical treatment was 630 million, accounting for 12.6 percent of the total. The proportion was very high both in cities and the countryside (1998 data): 10 percent and 14 percent, respectively; 38 percent did not receive medical treatment due to economic difficulty, and 70 percent should have been hospitalized but were not because they could not afford it. In cities about 41 percent in the low-income group were not hospitalized when they should have been. Even in the high-income group, about 17.2 percent who should have been hospitalized never were. Additionally, about 160 million, or 13 percent of China's population, suffer from chronic diseases. These data include both those who were diagnosed as having a chronic disease and those who were not yet diagnosed.[18]

More than 80 percent of the Chinese population, or 1 billion people, did not have medical insurance as of the mid-2000s. More specifically, over the past ten years the number of people not covered by medical insurance increased from 900 million to 1 billion, rising from 67.8 percent to 80.7 percent of the population. The figure for cities increased from 96.53 million to 300.00 million, with the proportion rising from 29 percent to 52 percent. This was due in large part to the increase in the urban population. The figure for rural areas dropped from 800 million to 400 million, due to the decline in the rural population during this period. But those not covered by medical insurance rose from 94 percent to 97 percent in rural areas. Data released by the National Bureau of Statistics show that the number of people covered by basic medical insurance in 1998 was 18.78 million. This figure rose to 110 million by 2003, accounting for 20.8 percent of the total urban population (523.76 million), or 8.4 percent of the total national population (1.29 billion). The number of urban people not covered by basic medical insurance rose from 397.30 million to 414.74 million, while the proportion dropped from 95.5 to 79.2 percent.[19] Even this figure is 110 million larger than the statistic obtained from a survey conducted by the Ministry of Health.[20] This third national health survey shows that 300 million to 400 million urban citizens are not covered by medical insurance, which does not include any of the 800 million people living in the rural areas.

The problems posed by ill health among women, children, and impoverished populations are very serious. Despite major progress in maternal and child health care, these groups still face serious health risks. Although the urban prenatal examination rate rose from 70 percent in 1993 to 88 percent in 2003 and the rural prenatal examination rate rose from 60 percent to 86 percent in the same years, far too many Chinese women still fail to obtain examinations. In rural areas an estimated 1.05 million women did not complete a prenatal examination by 2003. Then there is the problem of hospital birth delivery. The percentage of women giving birth in hospitals has reached 93 percent in cities, but the figure for the rural areas is only 62 percent. An estimated 2.86 million rural women give birth outside of hospitals each year.

The health of newborns has improved greatly. The proportion of infants who were underweight was only 3.1 percent in cities and 3.8 percent in rural areas in 2003, representing a relatively small rural-urban divide.[21] Both of these figures are lower than the world average of 16.8 percent, the lower-middle-income-country average of 9 percent, as well as the upper-middle-income-country average of 8 percent. The scheduled immunization program for one-year-olds is also good, better than in the average middle-income country, but still records show only 56 percent immunized one-year-olds in rural areas. An estimated 2 million children do not have immunization records.

China still deals with serious issues related to income poverty and health poverty. According to the Ministry of Health's *Third National Health Survey*, in the category of rural areas in which poor areas constitute one-third of the total rural population, per capita income was 1,183 yuan (743 yuan in real terms), significantly less than ten years earlier, when it was 922 yuan. In these areas about 60 percent of children do not have access to routine health check-ups, nearly half of pregnant women do not have pre- and postnatal examinations, and about 90 percent of women deliver their babies at home.

The percentage of households without access to clean drinking water is 5.3 percent, with 91.3 percent lacking sanitary latrines.[22] There is also a big gap between the cities and the countryside in terms of the infant mortality rate, the mortality rate of children under five, and the mortality rate of pregnant women. For example, as of 2008 the urban infant mortality rate was 0.65 per thousand and rural infant mortality rate was 1.84 per thousand; the urban mortality rate for children under five was 0.79 per thousand, and the rural mortality rate for children under five was 2.27 per thousand; the mortality rate for pregnant women in urban areas was 29.2 per a hundred thousand and in rural areas, 36.1 per a hundred thousand.[23] At present Chinese cities can match developed countries in terms of health levels, with a life expectancy of seventy-three years in 2008, although rural areas still lag far behind.[24]

China's tobacco consumption is the highest in the world, and smoking is one of the biggest obstacles to China's health. In the mid-1990s one out of three adults was a smoker, with the total number reaching 1.1 billion in the world.[25] Current surveys indicate that 47.3 percent of Chinese men smoke, a significant decrease from ten years ago (61.2 percent) but still considerably higher than the world average of one-third of men. In 1993 there were 270 million male smokers in China. By 2003 that number had dropped to 250 million, about 22.7 percent of the world total. Currently, 260 million Chinese citizens are smokers, or 26 percent of the adult population. This constitutes 23.6 percent of the global population of smokers. The average Chinese smoker smokes 15.9 cigarettes a day, totaling 5,800 a year. Of the fifty-eight countries whose data were included in World Health Organization statistics, only two exceeded an annual per capita cigarette consumption rate of 5,000: China and Iraq. China's 260 million smokers consume 1.5 trillion cigarettes a year, and in 2003 its cigarette production was 35.81 million boxes.[26] This production makes the tobacco industry one of the main sources of state revenue, but the long-run ill effects make smoking a serious obstacle to a healthy China.

China is also a liquor-consuming country, with a direct consumption of about 10 million tons a year. The percentage of alcoholics aged fifteen and up is 8.2 percent of the total population, or about 82 million.[27] In 2003 per capita liquor consumption in cities and the countryside was 9.39 kilograms and 7.67 kilograms, respectively. Annual consumption in these two areas totaled 4.92 million tons and 5.89 million tons.[28] Like the tobacco industry, the liquor industry is one of the main sources of state revenue, but excessive consumption is also a serious impediment to the health of the Chinese people.

Approximately 200 million to 300 million people in China do not have access to clean drinking water.[29] According to a 2005 survey, 34 percent of rural residents drank water deemed unsafe.[30] This would mean that roughly 300 million people do not have access to safe drinking water, with 191 million of them drinking water that contains harmful elements in a proportion that exceeds state standards. In rural areas in recent years more than fifty diseases were spread through drinking water. Unsafe drinking water is the main source of water-born endemic diseases and schistosomiasis. About 63 million residents of the countryside drink water with high fluorine levels, and about 38 million people drink bitter or salty water. Most of these people live in the north and east coastal areas. Schistosomiasis is prevalent in seven provinces, encompassing 110 counties and districts where about 60 million people live.

The threat of nontraditional major diseases is on the rise. The population structure and other factors have allowed chronic diseases such as cardiovascular and cerebrovascular illnesses, tumors, and diabetes to increase. The cases of tuberculosis in China made up 17.4 percent of the world total in 2008.[31] The *China Statistical Yearbook 2010* reports that in urban areas malignant tumors, heart disease, cerebrovascular disease, and diseases of the respiratory system are the leading causes of death, accounting for 78.68 percent of all deaths in 2009. In rural areas, malignant tumors, cerebrovascular disease, heart diseases, and diseases of the respiratory system were the first, second, third, and fourth most common causes of death, respectively, accounting for 79.62 percent of the total.[32]

Economic globalization and the resulting changes to the environment and people's behavior has led to several new communicable diseases becoming major threats to human welfare. Since the discovery of the first AIDS case in the United States in the 1980s, the disease has spread throughout the world. In recent years AIDS has spread rapidly across China. In 2005 the Chinese Ministry of Health estimated that there were between 540,000 and 760,000 cases of HIV and AIDS in the PRC, 65,000–85,000 of which were AIDS.[33] In 2005 alone, between 60,000 and 80,000 individuals contracted HIV, and 20,000 to 30,000 people died of AIDS. The *2008 Report on the Global AIDS Epidemic* states that there is a danger of rapid increases in AIDS cases in China, as there were then already about 700,000 HIV cases, rising from 490,000 in 2001.[34] The *2002 Report on the Global AIDS Epidemic*, published by WHO and UNAIDS, had warned that if effective countermeasures were not adopted quickly the number of HIV cases could increase tenfold over ten years, topping 10 million by 2010. The number is now less than 800,000, but this is still significant. One study estimates that there were between 600,000 and 1.3 million HIV cases in China in 2002 and that the cases were increasing at an annual rate of 20–30 percent. Extrapolating from these estimates, there will be between 11 million to 80 million cases of AIDS by 2015.[35]

Building a Healthy China by 2020

In August 2000 the UN Millennium Summit adopted the Millennium Declaration, establishing a series of Millennium Development Goals designed to eradicate human poverty and improve education and health through the concerted efforts of all countries (table 4-1).

The quantified goals with regard to poverty and health include

—Reducing by half, between 1990 and 2015, the proportion of people whose income is less than a dollar a day.

—Reducing by half, between 1990 and 2015, the proportion of the population without access to safe drinking water and basic sanitation.

—Reducing by three-quarters, between 2000 and 2020, the mortality rate of pregnant women.

—Reducing by two-thirds, between 1990 and 2015, the mortality rate of children under five.

—Reducing by half, between 2000 and 2020, the infant mortality rate.

China is a party to the Millennium Declaration and has committed to realizing these goals.

Since the 1990s Chinese progress on income indicators has exceeded Millennium Declaration targets, but it has been slower in making progress on its health goals. All of China's health indicators, except the pregnant women

Table 4-1. *Millennium Development Goals, China and Other Countries, 2007*
Units as indicated

Goal	High-income countries	Upper middle-income countries	China	China's 2020 goal
Health resources (per thousand persons)				
Number of doctors	2.7	2.2	1.4	2.0
Number of hospital beds	5.9	4.2	2.2	4.0
Public health and basic medical services (percent)				
Children immunized against measles	93	94	94	98
Prenatal medical checkup coverage	n.a.	93	90	95
Rural population's access to safe drinking water	98	84	81	90
Use of sanitary latrines in rural areas	99	61	59	80
Funds raised for equalizing health services (percent)				
Urban medical insurance coverage	n.a.	n.a.	50	100
Rural medical insurance coverage	n.a.	n.a.	99	100
Government spending on health as share of total spending	17	10	4	8

Source: UN in China, *China's Progress toward the Millennium Development Goals, 2008 Report;* Hu Angang, *China in 2020: Building up a Well-Off Society* (Tsinghua University Press, 2007).
n.a. = Not available.

mortality rate, have fallen short of the declaration's goals. The main reason for this slow progress are regional disparities in delivery of public health services. As discussed above, the difference between cities and the countryside has made it difficult for residents of impoverished rural areas to obtain basic health services.

Disparities in health services in China mainly manifest themselves in the urban-rural divide. Regrettably, China has not yet realized its goal of providing basic health care for all. Indeed, a portion of the underprivileged population does not have access to even basic health services. About 80 percent of China's health resources are concentrated in cities, with large hospitals accounting for two-thirds of the total. The amount of high-end medical equipment in large cities has reached or surpassed developed country levels. In fact, many cities have a surplus of such equipment. In contrast, public health organizations in small cities and counties, especially rural areas, lack basic medical equipment. The conditions are even worse in especially impoverished areas, where medical resources are scarce, houses tend to be in disrepair, medical equipment is out-dated and aging, and the quality of medical personnel is low.[36]

This gap is continually widening because the PRC lacks an effective transfer-of-payments system to balance the medical and health levels among regions. In other words, areas that are lagging behind economically cannot muster the funds necessary to ameliorate the serious inadequacies in their basic public health services. This has led to an extreme imbalance in the quality of public health in various regions. Intraprovince imbalances are also quite serious. Despite the fact that China has introduced a revenue-sharing system and the central government has strengthened transfer payments, imbalances remain acute. China's health development goals for 2020 are the following:

—Raise the mean life expectancy at birth from seventy-two years (2005) to seventy-six years.

—Reduce the infant mortality rate by one-third (from 2.33 per thousand in 2005 to 1.55 per thousand).

—Reduce the mortality rate of pregnant women by one-third (from forty-five per hundred thousand in 2005 to thirty per hundred thousand).

—Halt the spread of AIDS and other communicable diseases.

—Bring the rural population under the primary health care social security system.

—Raise the percentage of the population using tap water from 61.7 percent (2005) to over 90 percent.

—Raise the percentage of sanitary latrines in rural areas from 53.1 percent (2005) to over 90 percent.

—Solve the problem of more than 300 million rural residents (80 percent of whom reside in the middle and western parts of the country) lacking access to safe drinking water.[37]

—Increase the percentage of government investment in public health and basic medical services to about 7 percent of GDP.[38]

The objective of these goals is to build a "universally healthy" and "universally fit" society.[39] This can be achieved only if public health services and basic medical services are strengthened and the health levels of all people are improved. The following ten measures can aid in this process:

—Establish a basic health care system that covers residents in both cities and the countryside.

—Carry out a patriotic health campaign.

—Improve the emergency mechanism for handling public health emergencies.

—Build a disease prevention and control system.

—Build a public health and personal health information system.

—Improve the medical and health service system, with emphasis on the three-tiered rural health service networks and new community health service systems in cities.

—Develop a diversified medical service market to satisfy the demand of different groups of people.

—Establish a "fitness for all" service and management system and form a sports network under the guidance of the government and with extensive public participation.

—Increase fitness training stations to popularize sports and fitness among the people and increase the number of people participating in athletics.

—Encourage all people to exercise every day and make such activities more scientific, popular, and accessible.

Human Development Index and China's Goal of Building a Well-Off Society

The People's Republic of China recently celebrated its sixtieth anniversary. These sixty years are only a blink of the eye in the Chinese civilization's five-thousand-year history. Taking a step back to consider the Chinese civilization's evolution, one wonders what, in the world's most populous country that was also once its poorest, has been its biggest change. What miracles has China been responsible for in the history of mankind? Contemporary China's greatest accomplishment, I think, is to have transformed from a country of 540 million people with an extremely low human development level to a country of 1.3 billion people with an HDI on par with the world average. Under the leadership of the Chinese Communist Party, it took China only sixty years to make this dramatic progress in human development. This represents China's biggest contribution to world development and should be considered the greatest miracle in the history of human development.

CHANGES IN CHINA'S HUMAN DEVELOPMENT LEVEL

In its *Human Development Report,* the UN Development Program defines HDI scores below 0.50 as low, HDI between 0.50 and 0.79 as middle, and HDI between 0.80 and 1.00 as high.[40] China's HDI was very low when the PRC was established. In fact, in 1950 its score of 0.159, slightly lower than India's 0.160, gave it the distinction of being the least developed country in the world. At that time, China's per capita GDP was only $439 (PPP, 1990 international dollars), far lower than that of India ($619) and only 9.6 percent of the global average ($4,594).[41] China's HDI rose significantly from 1950 to 1975, steadily gaining ground on the developing countries, narrowing the HDI gap to the point where it was smaller than the economic development gap (PPP, per capita GDP). By 1975 China's HDI had risen to 0.523, higher than that of India (0.412). Despite China's tremendous HDI growth, however, it remained relatively undeveloped.

In the thirty years since reform and opening, major progress has been made in China's level of human development. China's HDI has risen 0.249 points, from 0.523 in 1975 to 0.793 in 2008. In terms of its HDI ranking worldwide, China and Bangladesh rose the fastest over the past dozen years.[42] In 2003 China was ranked in ninety-second place among the 182 countries listed, having advanced thirteen places since 1990 and nineteen places since 1992.[43] If HDI is calculated using the unified data source method, China ranked in eighty-eighth place among 136 countries in 1990 and rose to eighty-ninth place among 169 countries by 2008.

In 2008 China's per capita GNI was $7,259 (PPP), about 68 percent of the world average. In 2007 its HDI score of 0.772 was higher than both the average of developing countries and the average of medium-level HDI countries (0.686) and had begun to rise past the world average of 0.753. Among HDI indicators, the mean life expectancy at birth was 72.9 years, the adult literacy rate was 93.3 percent, and the gross enrollment rate of the three-tier education system was 68.7 percent. All of these measures exceeded the world average.

Between 1980 and 2007 the contribution to HDI from improvements to life expectancy, which rose by 0.125, was 19.6 percent; education, which rose by 0.21, contributed 33.0 percent; and GDP, which rose by 0.304, contributed 47.4 percent. In sum, economic growth is the biggest contributor to HDI, followed by education and health. In 2010, when the targets of the Eleventh Five-Year Plan are on course to be met, China's HDI will likely reach 0.800, edging into the ranks of the world's high-HDI nations. By 2020 China's HDI will rise to 0.878 (table 4-2), with economic growth remaining the top contributor to human development, followed by education and health.

Regional Variation of HDI in China

The different human development levels found in China's several regions form what is known as one China, four worlds. Over the past thirty years fundamental changes have taken place in the regional pattern of human development, with the third and fourth worlds moving toward the second and first worlds. In 1982 no area had attained the HDI level of the first world (HDI > 0.8). By 2006 eight provinces and municipalities (including Shanghai),

Table 4-2. *Indexes of Human Development, Life Expectancy, Education, and GDP, China, Selected Years, 1950–2020*

Year	Human development index	Life expectancy index	Education index	GDP index
1950	0.225	n.a.	n.a.	n.a.
1975	0.523	n.a.	n.a.	n.a.
1980	0.559	0.674	0.641	0.361
1985	0.595	0.686	0.663	0.435
1990	0.634	0.712	0.711	0.481
1995	0.691	0.742	0.758	0.571
2000	0.732	0.771	0.791	0.633
2005	0.777	0.792	0.837	0.703
2008	0.793	0.799	0.851	0.665
2010	0.800	n.a.	n.a.	n.a.
2020	0.878	n.a.	n.a.	n.a.

Source: Nicholas Crafts, "The Human Development Index, 1870 1999: Some Revised Estimates," *European Review of Economic History* 6, no. 3 (2002): 395–405; UN Development Program, *Human Development Report 2002,* p. 246; *China Human Development Report 2007/08; China Human Development Report 2009/10;* author's estimates.

n.a. = Not available.

whose combined population constitutes nearly one-third of the country, had entered the first world. In 1982 only Shanghai and Beijing had entered the second world. By 1990 however, twelve provinces had entered the second world, and all but Tibet had done so by 2006. This shift was highlighted by an increase in the percentage of the population living in the second world from 2.10 percent to 69.44 percent. In 1982 most provinces and autonomous regions were in the third and fourth worlds, with populations accounting for 97.9 percent of the country. By 1990 the percentage dropped to 62.55 percent, with only Tibet still remaining in the fourth world. By 2006 Tibet had also moved into the third world, marking a new leap in China's human development level.

It took China less than thirty years to realize such fundamental changes, averaging a leap in HDI for one-third of the population every ten years. By 2020 most of the Chinese provinces and 80 percent of the population will likely have attained first-world HDI levels, leaving less than 20 percent of the population in the second world. At that point, China will have achieved its goal of a well-off level of living for its people.

Table 4-3. *Human Development Index and Generalized Human Development Index, China and India, Selected Years, 1950–2020*

Index and country		1950	1975	2000	2003	2007	2020 (estimate)
Human development	China	0.225	0.521	0.718	0.755	0.772	0.878
	India	0.247	0.416	0.571	0.602	0.612	0.700
Generalized human development	China	1.230	4.770	9.020	9.730	10.250	12.700
	India	0.890	2.530	6.100	6.410	7.130	9.370

Source: Nicholas Crafts, "The Human Development Index, 1870–1999: Some Revised Estimates," *European Review of Economic History* 6, no. 3 (2002): 395–405; UN Development Program, *Human Development Report 2009;* author's estimates.

GENERALIZED HUMAN DEVELOPMENT INDEX

The HDI indicates the development level of a country or region. It can be used as an indicator to reflect the quality of a population, just like per capita GDP. Just as the economic aggregate of a country or region is the product of per capita GDP multiplied by total population, the Generalized Human Development Index (GHDI) is the product of HDI multiplied by total population. This measure indicates the total welfare of human development. As a measure that incorporates both human development and population size, the GHDI is an overall indicator of the quality and proportion of the population.

We may use this indicator to compare China and India in human development. In terms of its working-age population, China's population dividend period had a high peak value, came earlier, and lasted a short time, while India's had a low peak value, came later, but will last longer. If current birth control policies remain unchanged, huge changes will take place in the population distribution of China and India after 2030. India will overtake China in total population and working-age population, with the total population likely to be 200 million more than in China and the working-age population likely to be 220 million more than in China. Currently, China's GHDI is about 1.4 times larger than India's. With respect to total population, China and India have been converging since 1975, dropping from a difference of 1.51 times in 1975 to 1.21 times by 2000 and 1.08 times by 2020 (table 4-3). China's population is likely to be less than that of India by 2040. When viewed from the perspective of the GHDI, the gap between the two countries rose from 1.39 times in 1950 to 1.89 times by 1975 before dropping to 1.48 times by 2000 and 1.36 times by 2020. It is important to note that although China has been leading India in terms of HDI, the shrinking gap in population between the two has led to a convergence in GHDI.

To pursue comprehensive human development, it is imperative for China to maintain GHDI growth. With its HDI set to reach developed country levels, China should change its strategy so that both HDI and population size drive GHDI growth instead of HDI alone. China should continue to intensify its investment in human capital, improve the average educational level of its people, and properly adjust the age structure of the population in order to increase the growth rate of the child population and make sure that the absolute number of children does not drop significantly.

China has made major progress in terms of human development. Its GHDI now exceeds the world average, but with a large population and vast territory, serious regional disparities remain. China not only suffers from large development gaps between cities and the countryside but also faces regional development imbalances and a large number of people living in absolute poverty. In addition, social development lags behind economic development. With this in mind, China must work hard to realize a harmonious social and economic development process and resolve the problem of uneven growth.

CHAPTER FIVE

EDUCATION AND HUMAN RESOURCES

E ducation is fundamental for developing human resources and promoting the all-around development of a people. It is the cornerstone of national development and a fundamental pillar of the reinvigoration of the Chinese nation. Indeed, differences across nations in technological level and speed of economic development are chiefly the result of differences in educational levels. "Formal schooling fosters attributes in a population that are conducive to the acquisition of modern technology."[1] Knowledge products and human capital have so-called externalities or spillover effects. Because investing in people produces the biggest returns and the best results, it is therefore a profitable undertaking. Human capital mainly refers to the stock of competencies, knowledge, and personality attributes embodied in the ability to perform labor that produces economic value. Generally, human capital can be calculated as a population multiplied by education level. Only when education flourishes is it possible for human capital to increase, an economy to develop, and national power to expand. Latecomers to modernization usually prioritize educational parity over economic parity with leading countries. In this chapter I calculate human capital as the population over fifteen years old multiplied by its average years of education.

Since the founding of the PRC in 1949 China has gone from a big power full of illiterates to a big power with universal primary education (1949–77) and subsequently from a big power with universal primary education to a human resources powerhouse (1978–2000). Throughout this process, the PRC has turned its huge population from a burden into a significant advantage.

China's Educational Development and Setbacks, 1949–77

In 1949 China was the country with not only the largest population in the world but also the most illiterate and semi-illiterate (80 percent of its population). Given this, China was also extremely short on modern human capital. The number of university graduates stood at 185,000. The number of middle school graduates was 4 million, only 0.74 percent of the total population.[2] According to my calculations, the average number of years of education received by people whose age exceeded fifteen in 1949 was only one year; the world average was about four years.[3] In 1950 China's 338 million people made up 21.6 percent of the world's population but only 5.5 percent of the world's human capital.[4]

After the founding of the PRC, the government placed education at the top of its agenda. Schools at all levels developed rapidly, with the number of students also increasing. By 1978 China had established a completely modern education system, consisting of preschool education, primary education, secondary education, higher education, and adult education. It had also increased the enrollment rate of school-age children (ages five to twenty-two) from 49.2 percent in 1950 to 95.5 percent by 1978.[5] Primary education basically became compulsory. The percentage of the population that was illiterate or semi-illiterate decreased significantly. By 1978 China ranked first in number of students in school (average number of years of education received by individuals over age fifteen was raised to nearly four years, four times the 1949 average and higher than India's average of about 2.6 years). In 1975 China's human capital had reached 17.5 percent of the world total and was first in the world (2.4 times greater than India's).[6] These prereform achievements provided a solid foundation of human capital for China's post-1978 economic takeoff.

It was not always smooth sailing for higher education during the Mao era. The process featured three distinct stages: 1949–59 was a period of high-speed development, when the number of students at universities rose from 117,000 to nearly 1 million; 1961–70 was a period of decline, when the number of university students dropped to only 48,000, less than in 1949; during 1970–76 the number of university students increased again, though only slightly. The 1960s decline was caused by the 1959–61 famine and exacerbated by the Cultural Revolution launched by Mao Zedong in June 1966. All schools closed, and university admittance was postponed for half a year, which eventually became four years. University admittance resumed only after 1970, when Mao proposed admitting students from worker, peasant, and soldier backgrounds. Admitting students through "recommendation by the common people, approval by the leadership and reviewed by schools" looked fair, but in reality the enrollment rate was very low, resulting in greater social injustice. Although the number of university students rose a little, it did not exceed the number enrolled in 1959 until after Mao's death in 1976. Secondary and specialized technical schools were similarly affected during the sixties and seventies.

It is estimated that during the ten years of the Cultural Revolution more than 1 million fewer university students and 2 million fewer secondary and specialized technical students were trained than would have been the case normally. Without a doubt, the political decisions of the period between 1966 and 1976 caused serious setbacks and major losses in the accumulation of human capital in China.[7]

Becoming a Human Resources Power, 1978–2000

After being purged several times during the Cultural Revolution, Deng Xiaoping was reinstated as a state and party leader in 1977. Deng was strongly aware of the importance of education, science, and technology in pursuing modernization. He himself took the post of vice premier to take charge of education, science, and technology. Among his first policy initiatives was one that restored the university entrance examinations and reinstated the postgraduate studies system. In 1978, 400,000 students were enrolled in college and 10,000 postgraduates began their training.[8] Deng also established bachelor's, master's, and doctoral degree systems. Students were openly selected to study abroad, with 860 sent overseas in 1978 alone.[9] All of these moves stimulated the development of higher education, thus laying an important institutional basis for the long-run accumulation of human capital.

In the 1980s the Chinese government began promoting the nine-year compulsory education system. On December 3, 1980, the CCP Central Committee and the State Council issued joint decisions on problems concerning compulsory primary education, vowing to make primary education compulsory in the 1980s and junior secondary education compulsory in places that had the means. On December 4, 1982, the Constitution of the People's Republic of China was amended to read, "The state undertakes to run all kinds of schools, make primary education compulsory, and develop secondary education, vocational education, and higher education, as well as preschool education." On August 16, 1983, the Ministry of Education issued Interim Regulations on the Basic Requirements for Making Primary Education Compulsory. Three

Table 5-1. *Average Years of Education Received by People Aged Fifteen and Over, China and Other Countries, Selected Years, 1950–2010*
Years of education

Year	China	Developing countries	Developed countries	World
1950	1.00	2.05	6.22	3.17
1960	2.00	2.55	6.81	3.65
1970	3.20	3.39	7.74	4.45
1980	5.33	4.28	8.82	5.29
1990	6.43	5.22	9.56	6.09
2000	7.85	6.15	10.65	6.98
2010	9.00	7.09	11.03	7.76

Source: Robert J. Barro and Jong-Wha Lee, "A New Data Set of Educational Attainment in the World, 1950–2010," Working Paper 15920 (Cambridge, Mass.: National Bureau of Economic Research, 2010); author's calculations.

years later, on April 12, 1986, the fourth session of the Sixth National People's Congress adopted the Education Law of the People's Republic of China, thus providing the legal basis for compulsory education.

During this period, the main task was to raise the general enrollment rate at junior secondary schools and to consolidate the enrollment rate gains of primary schools. As of 1990 the enrollment rate at primary schools had reached 97.8 percent, with the enrollment rate at junior secondary schools reaching 66.7 percent. By 2000 the enrollment rate at junior secondary schools was 88.6 percent, enrollment at senior secondary schools was 42.8 percent, and college enrollment was 5.56 million.[10]

Internationally, China's human capital in 1950 was half the world average and 85 percent less than the average of developed countries. As of 1980, however, the PRC had reached the world's average level. It continued to rise above the world average in 2010, significantly reducing its gap with the developed world (table 5-1). From 1950 to 2010 China's average annual growth in terms of human capital was 3.73 percent, much higher than the world average growth of 1.27 percent and the developed country average (0.81 percent).

According to research conducted by both Western and Chinese scholars, China's educational achievements contributed an average of 2.92 percentage points to annual GDP growth between 1979 and 2003, higher than the United States (2.01) and Japan (1.66) but lower than South Korea (4.18). Direct contribution by education was 0.85 percent, with a spillover effect of 2.07 percent.[11]

Table 5-2. *Educational and Human Resources Indicators, China,*
Selected Years, 1990–2020

Units as indicated

	1990	2005	2009	2015	2020
Senior secondary education					
Students at school (million)	n.a.	40.00	46.41	45.00	47.00
Students in vocational education (million)	0.22	n.a.	21.95	22.50	23.50
Gross enrollment rate (percent)	21.9	52.7	79.2	87.0	90.0
Higher education[a]					
Students at school (million)	3.82	24.00	29.79	33.50	35.50
Gross enrollment rate (percent)	3.4	21.0	24.2	36.0	40.0
Undergraduate at school (million)	1.97	8.49	14.06	15.20	16.20
Graduate at school (million)	0.09	0.98	1.41	1.70	2.00
Education received by newly recruited labor (years)	9.45	11.3	12.0	12.7	13.5
Continuing education on the job (million)	n.a.	n.a.	170	290	350
People with a higher education (million)	16	70	98	145	195

Source: *China Education Yearbook 2010;* author's calculations.
n.a. = Not available.
a. College and above.

Becoming a Global Human Resources Power, 2001–20

The essence of modernization is the modernization of men and women, or the improvement of human capital. For China the modernization of its population of over 1 billion requires intense investment in education and dramatic improvements to human capital. When people's human capital evolves from the low to the middle level, significant progress will be attained in terms of total human capital.

Entering the twenty-first century, China has accelerated the development of its senior secondary educational apparatus. The Tenth Five-Year Plan outlined a goal of raising the general enrollment rate at the senior secondary level to 60 percent by 2005; it fell short, reaching only 52.7 percent (table 5-2).[12] As a result, this task was taken up again during the Eleventh Five-Year Plan period. By 2008 the general enrollment rate had reached 74.0 percent, with the number of students in senior secondary schools reaching 45.46 million, including 20.57 million students in secondary vocational schools students, who accounted for 45.2 percent of the total.

In terms of higher education, the Tenth Five-Year Plan set the goal of raising the general enrollment to 15 percent by 2005. That goal was met by 2002. The population with access to higher education reached nearly 98 million by 2009, 543 times that of the 1949 figure (185,000). This achievement marked the transition of higher education in China from a system of elite education to a system of mass education. The number of students at regular institutions of higher education stood at 21 million, while those in other forms of college programs (part-time and correspondence courses) reached 11.80 million. During the transition, China surpassed the United States (18 million) to rank first in the world in terms of students in higher education. Indeed, the number of students in the country's higher education system likely accounts for over a quarter of the world's total (about 100 million, according to conservative estimates).[13] Nevertheless, the general enrollment rate of higher education was 24.2 percent, still lower than that of the United States. The number of postgraduate students reached 300,000 in 2000 and then surged to 1.40 million by 2009, exceeding the number in the United States.

China has also opened its system of higher education to the outside world, becoming the Asian country with the largest number of foreign students. The total number of foreign students studying in China in 2008 was 223,500, nearly 180 times the number studying in China in 1978 (1,236) and more than the number of Chinese students studying abroad (178,900). Among them, 80,000, or 35.8 percent, were pursuing degrees, with 6 percent of the total, or 13,500, receiving Chinese government scholarships. These students come from 189 countries and study in 592 institutions of higher learning.[14] The total number of Chinese students who studied abroad from 1978 to 2008 is 1.39 million.[15] The United States was the biggest market for foreign students, accounting for 20 percent of Chinese students studying abroad.[16] Britain made up 12 percent; Germany, 10 percent; France, 7 percent; Australia, 6 percent; and Japan, 5 percent.[17] There is still huge potential for China to absorb foreign students. After all, its share of students enrolled in institutions of higher education is far greater than that of the United States and other developed countries. It will likely soon be in the front ranks of the world.

In the twenty-first century China has begun to implement its "going-out education strategy." By December 2008 twenty-four universities had set up forty-two branch schools or learning centers abroad. In addition, the PRC has opened 305 Confucius Institutes or classes in seventy-eight countries, has signed agreements for mutual recognition of degrees with thirty-four countries, and has maintained cooperative relationships with more than forty major international education organizations.[18] Developing education,

Table 5-3. *Total Human Capital, China, Selected Years, 1950–2020*
Units as indicated

Item	1950	1960	1982	1990	2005	2020
Population aged 15–64						
Million	337.8	363.0	625.2	762.6	888.0	996.0
Share of total (percent)	62.0	56.3	61.5	66.7	70.2	69.6
Average years of education	1.0	2.0	5.3	6.4	8.5	10.0
Human capital						
Billion persons per year	0.3	0.7	3.3	4.9	7.5	10.0
Share of world total (percent)	5.5	9.0	21.1	23.6	25.8	25.2

Source: UN Department of Economic and Social Affairs, *World Population Prospects: The 2008 Revision;* National Bureau of Statistics of China, "Statistical Data of the 55 Years of New China," *China Statistical Yearbook 2006;* author's calculations.

especially higher education, in an open manner is important as China seeks to secure its standing in the world. A developed system of higher education also projects soft power capable of influencing the world.

The trajectory of China's development in the realm of education and human resources is similar to South Korea's in the late twentieth century. In the 1960s the gross enrollment rate of secondary education in South Korea ranked twenty-fourth among the OECD countries. By 2000 Korea had climbed to second place. During that period Korea's gross enrollment rate in higher education rose from twentieth place to first place.[19] The difference between China and South Korea is that China's population is 27.5 times larger. If China's education development reaches even a fraction of that attained by Korea, it will become the world's leading educational power.

The factors contributing to this accelerated growth of total human capital include not only the growth of the working-age population but also a significant rise in per capita human capital. In 1950 China's human capital was 338 million persons/year (table 5-3). It shot up to 3.332 billion persons/year by 1982, averaging an annual growth rate of 7.4 percent during this period. Such high levels of growth were made possible in part by the fact that human capital in 1950 was very low, with the average number of years of education received by individuals over age fifteen being only one year. By 2005, just over half a century later, human capital reached 7.548 billion persons/year, or 2.26 times the 1982 figure. From 1982 to 2005 average annual growth was a bit lower (3.6 percent), largely because the starting point in 1982 (5.33 years of education) was significantly higher than the 1950 baseline. Today, China may be regarded as a human resources power, which is its most significant advantage as it seeks ever-higher levels of development.

Based on an international comparison of relative human capital levels in 2002, I predicted that China would become a world power in terms of human capital by 2020.[20] In 2003 I joined the China Education and Human Resources Development Project, organized by the Ministry of Education. One of the major strategic objectives set by the group for the subsequent twenty years in education and human resources development was for China to build the world's largest "learning society," which would foster an environment of learning and sharing knowledge. Through the construction of such a society, the Chinese government would aim to convert the country's heavy population burden into a rich human resources advantage. This first development objective, which was accepted as a 2020 education development goal by the Chinese government, was to "turn a big human resource power (*renli ziyuan daguo*) in terms of merely the quantity of college graduates into a strong human resource power (*renli ziyuan qiangguo*) in terms of comparative advantage in both quantity and quality."[21] In 2007 the Seventeenth National Party Congress report codified this goal, officially advancing the slogan "Prioritizing education development and building a big human resources power."

In pursuit of this goal, the government is attempting to raise the gross enrollment rate at senior secondary institutions to 90 percent by 2020, putting the number of students in these schools at 47 million, or twice as many as in the United States by 2020. By then the gross enrollment rate of higher education is also expected to reach 40 percent. By 2008 the number of college-age individuals began to decrease, a trend that will continue at least until 2015. Despite this, the number of students in school will continue to grow over the coming decade, to 35.50 million, which will include 2 million graduate students. The average number of years of education received by new entrants to the labor market will increase from 12.0 years in 2009 to 13.5 years in 2020, approximately the 2005 level of the OECD member countries (see table 5-2). By 2020 China will be the world's largest power in terms of population that has received higher education. In 2020 nearly 200 million Chinese citizens will have college degrees, more than double the number that did in 2008 (84 million). This means that the number of college-educated Chinese nationals will nearly equal the working-age population of the United States (220 million) and will be larger than the total employed population of the United States (180 million).[22] Even so, only 20 percent of Chinese workers will have received a college education. This percentage must be raised even higher.

By 2020 the number of foreign students in China will double, to reach 500,000, accounting for about 10 percent of the world total of students studying in a different country. The average number of years of education received by people aged fifteen and over will reach 10.0 years, 1.5 years less than the 2005 OECD average (11.5 years). The country's population aged fifteen to sixty-four will be nearly 1 billion, and total human capital will reach 9.96 billion people/year, 29.5 times the 1950 total and 25.2 percent of the world total. In seventy years, China will have progressed from a nation full of illiterates to one defined primarily by its strength in human resources.

In its quest to develop into a strong human resources power, China is facing both significant development opportunities and challenges. In terms of opportunities, educational development will provide impetus for steadily accelerating economic growth. According to studies by Wang Xiaolu and others, from 1998 to 2007 the direct contribution to and spillover effect of education—that is, human capital—accounted for 2.2 percent of GDP.[23] Without a doubt, education has been one of the most important factors in the Chinese economy's rapid growth. Wang Xiaolu expects this to continue, forecasting an economic growth rate of 9.34 percent between 2008 and 2020, with the net contribution of education to GDP expansion reaching 2.4 percentage points.

Education also has a direct impact on China's HDI, responsible for about 30 percent of its score. Education's contribution to basic research and knowledge innovation will rise from 70 percent in 2007 to over 80 percent by 2020.[24] Further, the equalization of elementary public education, by providing children from different areas and from different families equal access to education, contributes greatly to the narrowing of income gaps among regions and families, and thus helping to construct a harmonious society.

How Should China Become the World's Leading Human Resources Power?

China still has many policy problems to tackle if it is to become the world's largest human resources power by 2020. Among the most pressing is the fact that the current supply of education falls far short of the people's huge demand in terms of quality, diversity, and individuality. China has basically solved the problem of the affordability of education, but students and their parents demand better schools. In providing an education system for the world's largest student body, China faces a scarcity of public education resources. Though China's population receiving education makes up about 20 percent of the world total, its 2007 GDP (exchange rate method) makes up only 5.9 percent of the world total, and its public education–related expenditures account for only 4.2 percent of the world total. This is because the percentage of GDP that China spends on public education (3.2 percent) is lower than the world average (4.5 percent).[25]

Moreover, the current education system and its preferred teaching methods are flawed. The one-sided pursuit of increased university enrollment rates does not necessarily accord with the demands of the employment market or international competition. This simplistic approach to educational development also fails to satisfy the diversity of needs, and there is a large gap between town and country in education development. There is also a big disparity in the level of compulsory education services across regions. The differences in educational opportunities available at different schools and among different groups of people represent a basic injustice. Furthermore, a considerable number of teachers are not high quality, directly affecting education outcomes. The gross enrollment rate of preschool educational programs and kindergartens is very low (only 47.3 percent in 2008). It is necessary, therefore, for China to take the following six steps.

First, it is imperative that the state accelerate the popularization of preschool education, aiming to raise enrollment in kindergartens to over 75 percent by 2020. It is also necessary to improve the quality of the nine-year compulsory education system. The junior middle school completion rate needs to surpass 85 percent, the gross enrollment rate at senior secondary schools should reach 90 percent, and the graduation rate of students must hit 95 percent. The comprehensive enrollment rate of the three-tier education system (elementary, secondary, and higher) should be raised from 66 percent (the 2007 level) to nearly 80 percent by 2020, closing the gap between China and developed countries. Active steps should be taken to develop diversified vocational education programs and introduce free education, a grant-in-aid system, and a part-time study system in rural areas.

Second, it is essential to improve the quality of teachers. All primary school teachers should have at least a two-year college education and all new teachers for junior middle schools and primary schools should have received regular university schooling. The number of teachers at senior secondary schools who have completed postgraduate studies also must increase. The percentage of teachers in regular institutions of higher learning with a master's degree should exceed 90 percent. The percentage of professors at key universities holding a doctorate should also exceed 90 percent, with 60 percent holding degrees from abroad. The government must undertake the training of a contingent of outstanding educators and establish the honorary title People's Educator.

A modern, super-broadband-based education information infrastructure and network should be built to cover all schools in both cities and the countryside so that quality education resources—especially quality teaching resources—can be shared. Schools should be run in such a way as to improve students' learning abilities, practical abilities, and innovative abilities. For university graduates, job skills, innovative abilities, and international competitiveness must be improved. It is necessary to build a number of world-class universities and academic departments with distinct Chinese characteristics. They will serve as bases of research, knowledge innovation, and scientific and technology development.

Third, equality in education should be made the most important policy goal. The basic public education system should be improved such that it covers all people in both the cities and the countryside, especially the school-age population. It also must ensure that the rights and interests of all people are protected and that all people have equal access to education. Ceaseless efforts should be made to narrow the gap in education between cities and the countryside and among different schools in different regions. The urban educational system, especially nine-year compulsory education and secondary vocational education, should be extended to cover the children of migrant workers. Education in minority areas must also be improved, and financial aid should be made available to children from poor families. The compulsory education and secondary vocational education system should be opened to all handicapped youths so as to improve their developmental power. In poverty-stricken areas, in ecologically fragile areas, and in the middle and western parts of the country, more large boarding schools (primary, junior secondary, senior secondary, and vocational) should be established to increase students' chances of pursuing higher education. This will improve their job skills and foster the transfer of the rural population, especially young people and their labor, to cities and developed areas.

Fourth, the educational system must be reformed, and a premium must be placed on institutional innovation. At all levels, the government must be clear about its respective functions in promoting educational development. A national infrastructure and public education platform should be established under the leadership and support of the central government. The central government should design development plans and then provide support in terms of policy, public finance, and human resources. The national education platform should be managed level by level, with localities as the main managers and the provincial governments taking general responsibility for the development of education in the areas under their jurisdiction. Governments at the prefectural and county levels should facilitate planning for the development of preschool education, primary education, secondary education, and vocational education. Schools must be given a certain measure of autonomy and be allowed to manage themselves. They should be granted legal status as independent organizations responsible for their own operations, management, and development—but also subject to public oversight. Nongovernmental organizations and individuals should be encouraged to open schools; the number of such schools, the number of teachers, the number of students, and the amount of investment they contribute to noncompulsory education should be raised significantly. Enterprises and different trades and services should be encouraged to open secondary and tertiary vocational schools independently or in cooperation with other groups. Support should continue to be given to public universities as they open independent colleges and continuing education institutions to satisfy the demand for education at different levels and from different quarters.

Fifth, both international and domestic education resources should be fully employed. Taking into consideration the existence of two markets, the domestic and the international, schools should be opened to the outside world. The number of graduate students being educated domestically should increase, and more doctoral candidates should be granted scholarships to pursue short-term stays at first-rate international universities and research organizations. Students should be encouraged to study abroad at their own expense, but certain services should be provided to them. Students graduating from foreign universities should be encouraged to return to serve the motherland. The number of such students who decide to return should be doubled. At the same time, Chinese scholars residing abroad should be encouraged to serve the state in different capacities. First-rate international universities should be encouraged to open study centers and institutes in China in the form of cooperatives or joint ventures. Leading domestic universities should be encouraged to increase the number of courses taught in foreign languages. Five million high-level international personnel with international vision, knowledge about international conventions, and experience in international competition should be trained. Efforts should be made to attract exchange students and properly raise the tuition for students at their own expense.[26] At the same time, more scholarships should be provided by the central government and local governments for students coming to China from developing countries.

Sixth, an investment system should be established in which the government is the main investor and social organizations provide supplementary support. The system should strive to raise the percentage of investment in education as a percentage of overall GDP from 4.8 percent in 2008 to over 7.0 percent by 2020. Of this, the national administrative expenses in education as a percentage of GDP should be raised from 3.4 percent in 2008 to over 5.0 percent.

We can see clearly that China has been transformed from a largely illiterate country to one with very large human capital resources and that it is now on its way to becoming a country with strong and competitive human capital. Unprecedented progress has occurred in the domain of education since the founding of the PRC, especially after reform and opening. China has clear goals for its education system. The pursuit of these goals will be a significant factor in China's quest to become a superpower.

CHAPTER SIX

SCIENCE, TECHNOLOGY, AND INNOVATION

Today's China is much different from China circa 1978. The country's rapid reemergence and renaissance has propelled the PRC into a new era. Several centuries ago the European renaissance was ushered in by a proliferation of science and culture. The use of printing technology to spread ideas and knowledge led to social change, urbanization, globalization, technical revolution, and eventually the Industrial Revolution. The same can be said of China's modern renaissance, which has unfolded as the country's population has expanded. China's renaissance is characterized by several important trends. They include the accelerated absorption of scientific knowledge (mainly derived from North America, Western Europe, and Japan); the provision of strong incentives for indigenous Chinese innovation; the extensive utilization of new technology such as mobile phones, computers, the Internet, and broadband to efficiently spread knowledge, technology, information, and culture; and the PRC's open participation in economic globalization and international competition.

China entered into a golden age not only in economic development but also in the realms of science and technology.[1] The country has not only become the biggest modern success story in economic development but has also created a miracle in the development of science and technology, transforming the country into a strong innovative power in these domains.[2] By 2020 China will be an innovative country and the largest knowledge-based society in the world, certain to be making major contributions to human development.

Catching up in Science and Technology

How does knowledge reconfigure human society? What will China look like sixty years from now? One must keep these questions in mind when considering China's modern-day renaissance.[3] Creativity, innovation, and invention in the fields of science and technology (S&T) have transformed Chinese society from a stagnant and lagging traditional agricultural society to one transitioning toward a knowledge-based economy with impressive achievements in innovation.

Of the factors influencing economic development globally, knowledge is playing an increasingly important role. Numerous studies of knowledge as a mechanism for economic growth have led to the construction of new economic growth theories and a series of endogenous growth models. Knowledge has become the most important factor in interpreting the disparities of economic growth among countries and regions.[4] In the knowledge-based twenty-first-century economic race, knowledge will no doubt continue to be central. Knowledge resources are, therefore, the most important resource for promoting economic development in a given country or region. Conversely, low levels of knowledge resources will inhibit the successful development of a country or region.[5]

The world is composed of more than 200 countries. Among them, 11 have populations of at least 100 million, but only two have at least a million scientists and engineers engaged in research and development (R&D). The first is China (1.82 million in 2009) and the second is the United States (1.41 million in 2007).[6] Similarly, China and the United States are the only two countries with a university-educated science and engineering labor force that exceeds 10 million: in 2009 China had 20 million such workers, while in 2007 the United States had 16.6 million.[7] China's human capital in the realm of science and technology will allow it to become one the world's largest innovative, or knowledge-based, societies.

Power in Science and Technology and Methods of Measurement

Against the backdrop of the increasingly fierce global economic competition and the globalization of science and technology, improving one's strength in S&T is not only a development opportunity but also an unavoidable challenge. Every country must accrue power in S&T. No advance means retreat, and so does a slow advance.[8]

But what exactly is S&T power? How do we define it, and what are its implications? It is a concept that is difficult both to define and to analyze quantitatively. In 2002 my colleague Men Honghua of the Central Party School of the CCP and I studied the comprehensive national power of five large countries: China, India, Japan, Russia, and the United States. We defined comprehensive national power as the ability to pursue a strategic goal through purposeful action. Our definition of comprehensive national power included eight strategic resources and twenty-three indicators, including the highly weighted factor science and technology.[9] Other scholars, such as Jong-Hak Eun of Kookmin University, assert that a country's S&T power is determined by its reserve of scientific and technological knowledge and its ability to develop its S&T reserves for military and economic purposes.[10]

The following guidelines shape our understanding of the relationship between a nation's S&T resources and the pursuit and realization of its strategic development goals:[11]

—S&T power does not refer narrowly to the level of some particular technology as compared to the world. Rather, it refers to a country's general standing in S&T competition. That is to say, we must know whether a country's total power is rising or falling.

—S&T power does not refer to static power. Rather, it refers to a country's dynamic power, which must be assessed from both a historical and a future-oriented perspective in order to know whether it is rising or falling.

—S&T power does not necessarily refer to a country's absolute power. Rather, it refers to its relative power—its S&T power relative to that of other countries and its proportion of the world total.

—Given economic globalization and given that the globalization of S&T is the most important factor of production—as well as the central importance of knowledge and information—S&T will inevitably move across the globe. With this in mind, a country's S&T power is not confined to the utilization of these resources domestically. Instead, it must include the country's ability to use them on a global scale. The more open a country is, the greater is its opportunity to receive an inflow of S&T resources. Openness also increases a country's ability to use its S&T resources globally, thereby strengthening its overall S&T power.

Men Honghua and I define the S&T power of a country as the sum of its ability to obtain, utilize, and allocate its S&T resources. It is made up of five capacities. We used five major indicators to compute these capacities for use in quantitative analysis. The five capacities and five indicators cover the most important and most crucial information but are not necessarily all-encompassing.

The first capacity is the country's innovative capacity in science. This is measured by calculating the number of papers published by that country's citizens in the more than four thousand international academic publications that cover such areas as physics, biology, chemistry, mathematics, medicine, engineering technology, and space technology. These publications represent the innovative power of a country and its influence in the world of international science. The more papers published in international academic publications, the greater the impact that country has in the realm of international science and technology. Here, quantity comes first and quality second.

The second capacity is innovative capacity as regards technology. This is measured by calculating the number of patent applications filed by the residents of a country with their own patent offices. The number of invention patents filed in a country is indicative of the technical innovative strength of that country and its influence on the world's technical progress. The implication here is that technology produced domestically spills over into the international arena. Once it is applied by an enterprise, it sharpens the competitive edge of the country in which that enterprise resides.

The third capacity is the country's capacity to use new technologies. This is measured by calculating the number of people using computers in a given country. The number of people using computers, which are the most representative, universal, and popular modern form of technology, demonstrates the power of a country to use new technologies. Once computers are mastered by the people, a country's labor productivity and international competitiveness will increase.

The fourth capacity is the capacity of a country to use global information. This is measured by calculating the number of people that access the Internet. Accessing the Internet demonstrates the power of a country to obtain information and its influence on the world's information network. The Internet has the function of spreading and magnifying information. Once it is extensively used, it helps to enhance the information capital and knowledge capital of a given country.

The fifth capacity is the capacity of a country to invest in R&D, measured by the country's R&D expenditures (purchasing power parity, or PPP).[12] There are two methods for calculating these expenditures. The first uses the official, or nominal, exchange rate. The second uses PPP, or the amount of a certain basket of basic goods that can be bought in a given country with the money it produces. Sometimes it is called the international dollar. The

International Comparison Program used by the World Bank and the United Nations sets 1993 as the base year and uses PPP as the conversion factor in estimating the international dollar value of per capita GNP and GDP in 118 countries. This method of calculation greatly underestimates China's investment in R&D and overstates the investment of developed countries.[13] R&D expenditures represent both investment power with regard to R&D and the potential for further R&D.[14] The greater the input in R&D in a period, the greater the output of R&D in future periods.

Of the five indicators, the first two show the innovative capacity of a country as regards knowledge and technology, while the latter three reflect a country's use of knowledge and technology—that is, the economies of scale of science and technology. The first four indicators are both output indicators and physical indicators and can be compared across international borders. The last indicator is both an input and a value indicator. Indicators focused on output take precedence, truly reflecting the S&T power of a country and facilitating international comparisons.

Each of the five indicators is assigned a value. To make it simple, we use equal weights to calculate the S&T power of a country. The result is a figure for relative power, or the proportion of the aggregate world S&T power. If such a percentage has risen for at least twenty-five years we call it the rising type; if it has fallen for at least twenty-five years we call it the declining type. We may also use this measure to compare the relative gaps between different countries.

It is important to note that the amount of S&T power is not indicative of the level of S&T development. The former refers to the aggregate of the indicators (national gross indicators), while the latter refers to the quantitative index (per capita index). The former refers to the competitiveness of a country in the world, while the latter refers to the lead a country holds in the world's S&T development. A country with a high level of S&T development is not necessarily influential on the world stage. However, a country with a low level of S&T development may still have a big influence on global affairs.

China as a Science and Technology Power

The United States, Japan, Germany, the United Kingdom, and China can be regarded as the five S&T powers in the present-day world. Combined, they make up over 57 percent of the world's total of S&T power, although their total S&T workforce is only 28 percent of the world's. These five powers determine the course and orientation of S&T development in the world, as they make the biggest contributions. Of the five big powers, only China is

Table 6-1. *Science and Technology Papers Published Internationally, China, Selected Years, 1987–2008*

Rank

Year	Science Citation Index	Index to Scientific and Technical Proceedings	Engineering Index
1987	24	14	10
1990	15	13	9
1995	15	10	7
2000	8	8	3
2005	5 (5.3%)	4	2 (12.6%)
2008	2 (8.1%)	2 (12.5%)	1 (22.5%)
2009	2 (8.8%)	2 (12.8%)	1 (23.9%)

Source: Ministry of Science and Technology of the People's Republic of China, *S&T Statistics Data Book 2010.*

Note: Percent of the world total is shown in parentheses.

a developing country. Despite its status as a developing country, China has the world's largest population and boasts the largest reserve of S&T human resources. China used to be a "straggler" in the realms of S&T, and it joined the race only thirty years ago. In one generation, however, it has become a "rapid pursuer," having overtaken the United Kingdom and Germany in a number of indicators. Historically and internationally, China's performance is unprecedented indeed.

PAPERS IN INTERNATIONAL SCIENCE AND TECHNOLOGY JOURNALS

At the beginning of reform and opening China produced few papers for publication in international academic journals in science and technology. As of 1980 the number of papers written by Chinese scientists and technical personnel constituted only 0.33 percent of all papers published in international journals, 119 times less than the U.S. scholars and 23 times less than Japanese scholars. By 2008, however, China ranked second in the world.[15]

In 1987 China ranked twenty-fourth in the Science Citation Index (SCI) and tenth in the Engineering Index (EI), two prominent indexes of journals in those fields (table 6-1). As of 2008 China had advanced to second in the SCI, accounting for 8.1 percent of papers published. In the EI, China had risen all the way to first, accounting for 22.5 percent of papers. In the Index to Scientific and Technical Proceedings (ISTP), China ranked second only to the United States (29.2 percent), totaling 12.5 percent of papers. Overall, science and technology papers published by Chinese citizens in major international

Table 6-2. *Science and Technology Papers Published Internationally, Five Major Powers, Selected Years, 1981--2007*

Percent of world total

Country	1981	1985	1990	1995	2000	2003	2007
China	0.3	0.5	1.1	1.4	2.2	4.2	7.0
Japan	7.5	8.3	8.1	9.0	9.1	8.6	7.1
Germany	8.0	5.7	6.8	6.5	6.9	6.3	7.4
United Kingdom	9.2	9.1	8.2	7.8	7.8	6.9	8.4
United States	39.4	38.6	40.5	32.7	30.9	30.2	30.5
U.S./China gap	119.3	71.6	36.1	23.0	9.5	7.2	4.3

Source: World Bank, *World Development Indicators 2006;* National Bureau of Statistics of China, "Statistics and Analysis of Chinese Papers in S&T."

journals accounted for 11.5 percent of the world total, second only to the United States' 26.6 percent. This indicates that China has truly become one of the biggest innovators and producers in the domain of science and technology.

The number of Chinese papers published in the Social Sciences Citation Index in 2008 accounted for 1.6 percent of the world total, for a ranking of eleventh.[16] This indicates that China's international position in the areas of humanities and social sciences also rose quickly, even though lagging far behind its advances in the natural sciences.

In a cross-country comparison, the gap in science and technology between China and the United States is narrowing rapidly. In terms of the quantity of papers published in international academic publications, the percentage gap between the two countries narrowed from 119.3 in 1981 to 9.5 in 2000 and then to 4.3 in 2007 (table 6-2).

China has edged into the ranks of the advanced countries in some major research areas, especially in basic sciences such as physics and chemistry as well as applied sciences such as electronics, communications, materials, and computer science.[17] China's Ministry of Science and Technology has given high marks to the achievements of Chinese scientists, who are becoming the backbone of major institutions in several fields and promise to contribute to long-term economic growth.[18] For eleven disciplines—engineering, materials science, chemistry, mathematics, physics, computer science, multidisciplinary science, pharmacology and pharmacy, geology, environmental science, and space science—the number of Chinese papers exceeds 5 percent of the world total.[19]

Chinese papers on science and technology have not only increased in quantity but have also improved in quality. China has edged into the top ten in the world in terms of the quality of its papers. According to Social Sciences Citation Index data, from 1992 to 2001 Chinese papers were the nineteenth most cited in the world. This rank improved to thirteenth for the 1996–2005 period and then to tenth for 1998–2008.[20] According to the China Science and Technology Information Institute, in the period from 1996 to 2005 Chinese papers ranked within the top ten in six areas. From 1998 to 2008 China was ranked within the top ten in eleven disciplines: engineering, materials science, chemistry, mathematics, physics, computer science, multidisciplinary science, pharmacology and pharmacy, geology, environmental science, and space science.[21]

The history of science suggests that quantity precedes quality. China is still in the first stage of development but has already catapulted itself into the ranks of the elites. It is now moving to the second stage, edging itself into the top ten in terms of the quality of academic papers. The next goal, set by the Chinese government in 2006, is to move into the top five by 2020.[22] This objective reflects China's pursuit of continued innovation in the sciences as well as its step-by-step plan for the long-term development of science and technology.

I predict that despite the fact that Chinese scholars have to write in foreign languages to publish in international journals, within a generation China will become the number-one nation in the world in terms of the publication of research papers. Once it has reached this point, China will serve as a treasure trove of global knowledge, featuring the largest number of first-rate research papers.

PATENTS AT HOME AND ABROAD

China began preparations to establish a patent system in 1978 and opened the State Patent Office in 1980. In March 1984 the Standing Committee of the National People's Congress adopted the Patent Law of the People's Republic of China. It went into effect April 1, 1985. As of 1986 there were a limited number of patent applications and patents granted. According to the World Intellectual Property Organization data bank, China's State Patent Office had received only 8,009 invention patent applications as of 1986, 39.5 times less than the number filed in Japan and 15.1 times less than in the United States.[23]

Between 1986 and 2007, however, the number of yearly applications surged to an annual rate of 17.7 percent, far exceeding the corresponding application growth rates of the United States (6.5 percent), Japan (1.1 percent), Europe (6.0 percent), and Germany (1.7 percent) (table 6-3). Growth in Chinese invention patent applications even outpaced robust domestic GDP growth, which averaged 10.0 percent a year over the same period. This shows

Table 6-3. *Invention Patent Applications, Six Major Powers, Selected Years, 1986–2007*

Number, except as indicated

Country or country group	1986	1996	2007	Average annual growth (percent)
United States	120,916	211,946	456,154	6.5
Japan	316,162	376,674	396,291	1.1
China	8,009	22,742	245,161	17.7
South Korea	12,755	90,326	172,469	13.2
Europe	41,342	64,035	140,763	6.0
Germany	43,114	51,833	60,922	1.7
U.S./China gap	15.1	9.3	1.9	. . .

Source: World Intellectual Property Organization, *World Intellectual Property Indicators 2009* (Geneva), p. 17.

that economic growth was driven by technical innovation and that technical innovation became, to a large extent, an endogenous variable of economic growth, with the speed of innovation exceeding economic growth.

The rapid growth of invention patent applications in China has helped narrow the gap in technical innovation between the PRC and the developed world. The gap between China and Japan, which was a multiple of 62.9 in 1985, narrowed to 2.4 by 2007. Over that same period, the number of applications filed in the United States went from 14.6 times the number filed in China to 2.3 times the number (table 6-4). By 2008 the number of invention patent applications filed in China had increased another 38 percent, to 290,000, further narrowing the country's gaps with Japan and the United States. It took China only about twenty years to become the third-largest country in terms of invention patent applications, and by 2007 China was the fourth-largest country in terms of invention patent grants.[24] The goal set by the Chinese government was to edge into the first five in terms of invention patent grants by 2020, which was achieved thirteen years earlier.

Over the past few years the number of international patent applications has risen rapidly, with China's percentage of the world total rising from 1.39 percent in 2004 to 3.7 percent in 2008. This elevated the PRC to fifth in the world in terms of international patent applications. The director general of the World Intellectual Property Organization, Francis Gurry, claims that the rapid rise in PRC international patent applications is indicative of the steady growth of China's innovative capacity.[25] The establishment of the patent

Table 6-4. *Invention Patent Grants, Five Major Powers, Selected Years, 1980–2007*

Percent of world total

Country	1980	1985	1990	1995	2000	2007
China	n.a.	0.7	0.9	1.4	3.1	8.9
Japan	33.2	44.0	50.1	47.3	46.7	21.6
Germany	5.7	5.2	5.6	5.4	6.3	2.3
United Kingdom	3.9	3.2	3.0	2.6	2.7	0.8
United States	12.4	10.2	13.6	17.6	20.0	20.6
Japan/China gap	n.a.	67.6	56.9	33.3	15.1	2.4
U.S./China gap	n.a.	16.0	15.5	12.4	6.5	2.3

Source: World Intellectual Property Organization, *World Intellectual Property Indicators 2009* (Geneva), p. 37.

system in 1980 has greatly stimulated the growth of independent R&D capacity. It can be expected that, along with further increases in China's invention patent applications vis-à-vis the world total, the PRC may well become a global center of technical innovation. At present China is rapidly transforming from a technical imitator into a technical innovator.

On June 5, 2008, the State Council officially outlined a program for the state's intellectual property rights strategy. The goals set in the program include edging into the global elite in terms of the number of invention patent grants within five years, boosting the number of patent applications filed by Chinese citizens in other countries, cultivating a number of international name brands, significantly increasing the proportion of GDP composed of core copyright industry output, and by 2020 realizing a China featuring fairly high levels of intellectual property rights creation, application, protection, and management. The implementation of this national strategy will make China one of the biggest countries in terms of invention patent grants, on par with innovative powers such as the United States and Japan.

INVESTMENT IN R&D

China's investment in R&D as a proportion of GDP has steadily declined since 1978. At the outset of reform and opening 1.45 percent of GDP was invested in R&D. That share dropped to 1.14 percent in 1985, 0.75 percent in 1990, and then 0.57 percent in 1995, before climbing back to 0.9 percent in 2000 and 1.8 percent in 2010.[26] The decline was closely associated with the change in the percentage of state fiscal revenue in GDP, which was

Table 6-5. *Research and Development Expenditures, Five Major Powers, Selected Years, 1980–2007*

Percent of world total

Country	1980	1985	1990	1995	2000	2004	2007
China	2.1	2.3	1.7	2.5	4.7	7.5	8.1
Japan	9.0	10.0	11.5	11.6	9.4	9.4	10.0
Germany	6.9	6.2	6.0	5.2	5.3	4.5	4.8
United Kingdom	4.8	3.9	3.6	3.3	3.0	2.7	2.4
United States	26.0	27.1	25.1	25.3	26.6	24.3	23.8
U.S./China gap	12.4	11.8	14.9	10.0	5.7	3.2	2.9

Source: World Bank, *World Development Indicators 2010.*

31.1 percent in 1978, fell to 10.3 percent by 1995, and then rebounded to 13.5 percent in 2000 and 20.4 percent in 2009.[27] The similar trajectories of these two measures demonstrate the increasing absorptive capacity and input capacity of the state public finance. International comparisons show that China's R&D expenditures as a proportion of GDP exceed those of other rising powers such as India and Brazil. Indeed, China's R&D expenditure as a percentage of GDP is the highest among developing countries, though it is just slightly higher than the world average (1.6 percent) and the OECD average (2.2 percent).[28] The growth of China's R&D expenditures, however, is among the fastest in the world. During the period lasting from 1981 to 2004, China's R&D expenditures grew at an average annual rate of 9.9 percent (calculated using the 2000 international dollar price), higher than that of Japan, the United States, and other developed countries.[29] From 1995 to 2009 China's R&D expenditures grew at an average annual rate of 21.7 percent.

China's investment in R&D as a proportion of the world total was less than 2.1 percent in 1981. It rose to 2.3 percent by 1985 before dropping to its lowest point in 1990. It then rose steadily to 2.5 percent in 1995, 4.69 percent in 2000, and 8.1 percent in 2007 (table 6-5). The relative gap between China and the United States was estimated to be 12.4 times in 1981 and 14.9 times by 1990 before being narrowed to 2.9 times in 2007. According to OECD data, China's R&D expenditures in 2007 reached $86.8 billion, the third largest in the world after the United States ($368.8 billion) and Japan ($138.8 billion).[30] By 2010 China's R&D expenditures as a percentage of the world total are likely to reach about 10 percent, about two times less than that of the United States.

Table 6-6. *PC Users, Five Major Powers, Selected Years, 1990–2007*

Percent of world total

Country	1990	1995	2000	2004	2007
China	0.38	1.17	4.45	6.31	7.41
Japan	5.55	6.28	8.44	8.28	n.a.
Germany	4.88	6.08	5.97	5.61	5.15
United Kingdom	4.65	4.91	4.36	4.35	4.72
United States	40.67	35.91	34.76	26.64	24.11
U.S./China gap	107.03	30.69	7.81	4.22	3.25

Sources: World Bank, *World Development Indicators 2006; World Development Indicators 2009.*

It is important to note that even if R&D investments are calculated using the PPP method, the total may underestimate China's share of aggregate global investments. This is due to the fact that the cost of manpower in R&D in China is far lower than that cost in OECD countries. In addition, China's government allocations for R&D are tightly restricted, with most of them not permitted to be used on labor costs.

PERSONAL COMPUTERS

China's share of personal computers (PCs) as a proportion of the world's total was virtually zero in the 1980s. As of 1988 the relative gap between China and the United States in number of PC users was 180 times. But then the share increased to 0.38 percent at a time when the United States could boast a penetration rate of 40 percent, 107 times that of China in 1990.[31] This indicates that China is a latecomer in terms of the use of new information technology. Beginning in the 1990s, however, the number of PC users in China began to surge such that it overtook the United Kingdom and Germany, arriving at third place in 2004. In 2007 China overtook Japan, to rank second in the world, after the United States. That same year, the relative gap between China and the United States narrowed to 3.25 times (table 6-6).

China has also become a major global PC producer.[32] In 2000 China produced 6.72 million PCs. By 2009 that number had risen by a multiple of twenty-seven to 182 million. China's annual output of PCs in 2000 accounted for 19.2 percent of the world total. By 2005 that percentage had already risen to 83.5 percent. Since 2000 the use of PCs in the PRC has risen exponentially. In 2000 there were 9.7 PCs for every hundred urban households. As of 2008 the figure was 65.7. With PC costs dropping, quality and performance improving, and per capita incomes rising, rural families have begun

Table 6-7. *Internet Users, Five Major Powers, Selected Years, 1990–2008*

Percent of world total

Country	1990	1995	2000	2004	2007	2008
China	0.05	0.15	5.77	11.02	14.70	18.58
Japan	0.95	5.05	9.74	7.52	6.38	5.67
Germany	3.78	3.79	6.35	4.69	3.98	3.86
United Kingdom	1.89	2.78	4.04	4.28	2.96	3.04
United States	75.75	63.18	31.79	21.68	15.51	14.40
U.S./China gap	151.50	412.20	5.51	1.97	1.06	0.78

Source: World Bank, *World Development Indicators 2006; World Development Indicators 2009;* Central Intelligence Agency, *World Factbook 2010.*

to purchase PCs. In 2009 there were 7.5 PCs for every hundred rural households, a situation similar to that found in China's cities toward the end of the 1990s. In addition, all types of schools in both the cities and the countryside have begun to use computers.

INTERNET USE

China was introduced to the Internet in 1987. The relative gap in the number of Internet users between China and the United States in 1993 was of a factor of 3,000. This gap was narrowed to 412 times by 1995 and further to 5.5 times by 2000 and 1.06 times by 2007. By 2008 the number of Internet users in China had reached nearly 300 million, overtaking the United States (table 6-7). In 1990 Internet users in the United States accounted for three-quarters of the world total, while those of China represented a meager 0.05 percent. China's share rose to only 0.15 percent by 1995. After that, however, China's percentage of the world total rose rapidly, to reach 5.75 percent by 2000, overtaking the United Kingdom. By 2004 China's share exceeded that of Japan. Still, only 22.5 percent of the total population uses the Internet, meaning there remains great potential for growth in Internet use.

China started using broadband technology in 2000. In 2001 the relative gap in the number of broadband users between China and the United States was 37.7 times. By the end of 2009, however, China had 136 million broadband users, surpassing the number of broadband users in the United States in June 2009 (113 million).[33] Oliver Johnson, the CEO of the British online magazine *Point Topic,* calls it "a major milestone for China," noting that "launching people into space is spectacular, but having the biggest broadband market down here on earth means a lot more for building a modern, high-tech economy."[34] In the future, China will strive to develop super broadband so as to build the world's largest network.

From Straggler to Innovator

Thanks to the policies of reform and opening, China has transformed from a straggler to an innovator in science and technology. Indeed, its S&T power has been rising steadily for at least a generation, enabling the country to sustain rapid rates of development. Over the course of this generation, China has caught up to and subsequently overtaken the United Kingdom, Germany, and other Western European industrialized countries to become the third-largest S&T power in the world. In the process, the gap between the PRC and countries such as Japan and the United States in terms of S&T power has been narrowed. In the next dozen years China will accelerate its pursuit of these two countries. If China can overtake the United States, it will become a true superpower in the realm of science and technology.

For a long time, China lagged behind the world not only in industrialization and modernization but also in science and technology. In 1956 Mao Zedong, in his 1956 work "On Ten Major Relationships," described the basic situation of China as first "poor" and second "blank": "By 'poor' I mean we do not have much industry and our agriculture is underdeveloped. By 'blank' I mean we are like a blank sheet of paper and our cultural and scientific level is not high."[35] But Mao was not pessimistic about the country's future. Rather, he looked favorably at the country's blank state: "We are like a blank sheet of paper, which is good for writing on." In 1964 China exploded its first atomic bomb, and Mao Zedong stated that China had changed its poor S&T position in the world. He then set the goal of realizing the four modernizations, including science and technology, by the end of the twentieth century but did not provide clear targets for S&T modernization. The question remained, then, how China should realize its goals.

In this regard, Mao has provided some direction. Two years earlier, in 1964, in "Build China a Powerful Modern Socialist Country," he had pointed out that China should not follow the beaten track of other countries and slowly crawl behind them in the development of technology: "We must break the convention and adopt advanced technologies as much as possible and make our country a powerful modern socialist country in a not too long historical period." At the time, the Chinese leadership was not clear about China's level of science and technology and how big the gap was between the PRC and developed capitalist countries. The leadership did know, however, that the gap could only be reduced with the addition of advanced technologies. But how was China to obtain such technologies? By adopting those of developed countries (such as the old Soviet Union)? By self-dependence? Mao Zedong and Zhou Enlai did not provide an answer.

Regardless, China's realization of the four modernizations was derailed by the launch of the Cultural Revolution in 1966. Some leaders, such as CCP Politburo member Jiang Qing, held on to their narrow views on modernization like "frogs at the bottom of a well." At several CCP Politburo meetings in 1974, Jiang Qing took advantage of the *Fengqing* incident to attack Zhou Enlai and Deng Xiaoping for their "servility to things foreign and for their assertion that buying ships is better than building them and renting ships is better than buying them."[36] At the time, Jiang had no knowledge that Japan had built a 40,000-dwt oil tanker, while the *Fengqing* was merely a 10,000-dwt ocean-going vessel.

Before 1978 China was basically marginalized in the world of advanced science and technology. When Deng Xiaoping (in his capacity as vice premier of the State Council and head of the Chinese government delegation to attend the Sixth Special Session of the United Nations) visited New York and Paris in 1974, he realized how wide the gap was between China and developed countries.[37] At that point, he decided to make modernization of science and technology the most important development goal. He also called for learning from Japan and resorting to "take-overism" in developing science and technology.[38]

Based on my research, China's S&T power in 1980 accounted for 0.81 percent of the world total, far lower than China's percentage of the world population; it was also thirty-two times less than that of the United States and twenty times less than that of Japan (table 6-8). The size of the gap was due in large part to the fact that it had been accumulating over the course of a century. China was indeed lagging far behind in the world's S&T revolution. No one at the time could have foreseen that China could so quickly narrow the gap between it and developed countries; nor could one imagine that it would take China only one generation to become a strong S&T power.

In the period between 1980 and 1985 China's S&T power as a proportion of the global aggregate rose from 0.81 percent to 1.16 percent. During this time, the gap between the PRC and countries such as the United States and Japan decreased. Between 1985 and 1990, however, the percentage dropped back down to 0.82 percent, close to the 1980 level. This change was caused by a large decrease in the percentage of state financial receipts as a proportion of GDP. If calculated by the constant price, the absolute figure of government spending on science and technology was lower than in 1985, and

Table 6-8. *Science and Technology, Five Major Powers, Selected Years, 1980–2007*[a]

Percent of world total

Country	1980	1985	1990	1995	2000	2004	2007
China	0.81	1.16	0.82	1.33	3.97	7.21	9.70
Japan	16.54	20.76	15.24	15.85	16.31	14.50	14.57
Germany	6.88	5.71	5.41	5.39	6.10	5.26	5.22
United Kingdom	5.95	5.38	4.27	4.29	4.35	4.04	4.06
United States	25.93	25.32	39.13	34.92	28.59	24.50	23.24
Total	56.11	58.33	64.87	61.78	59.32	55.51	56.79
U.S./China gap	32.01	21.83	47.72	26.22	7.20	3.40	2.40
Japan/China gap	20.42	17.90	18.59	11.90	4.11	2.01	1.50

a. Weighted calculation based on data in tables 6-2 and 6-6.

its proportion of GDP therefore dropped accordingly. At the time, private enterprises were not major players in technical innovation. China had just established a patent system, and Chinese scientists had just begun to participate in international cooperation and exchange programs. Meanwhile, the United States and other developed countries had already initiated a revolution in information technology and moved far ahead of China in such new technologies as PCs, mobile phones, and the Internet.

At the beginning of the 1990s, China began to import such information technologies as the Internet and mobile phones and opened up the manufacturing and service markets, encouraging market competition, providing incentives for private Internet and communications technology consumption, and increasing government investments in related infrastructure. All of these initiatives brought China closer to the developed countries and closer to overtaking the United States. China's share of global S&T power rose from 0.82 percent in 1990 to 3.97 percent in 2000, making the PRC fifth in the world, already a true emerging S&T power. By 2007 China's share of global science and technology power was 9.7 percent, putting it ahead of Germany and the United Kingdom and behind only the United States and Japan, which remained, respectively, 2.4 times and 1.5 times greater than the PRC. The next step is to catch up with and overtake Japan and the United States to become the world's largest S&T power. According to my quantitative analysis, China is likely to attain this goal.

Driving Forces of China's Rise in Science and Technology

Why is it that it has taken China only one generation to enhance its S&T power and become a world leader in these fields? How has China leveraged its opportunity and attracted popular support? What factors have driven this process? Will they continue to increase in importance?

I hold that China has very effectively taken advantage of economic and knowledge globalization, the rise of all economies in Asia, and domestic political stability, as well as rapid economic development and social progress. This is a period of strategic opportunity that it has never before seen and that has ushered in a golden age in China's science and technology development.[39] It is dependent upon four major driving forces: economic and S&T globalization, the business sector, the Chinese government's guidance and promotion, and sustained high-speed economic growth. These factors are both interrelated and mutually reinforcing.

ECONOMIC AND S&T GLOBALIZATION

Globalization has brought about major changes in the development model of science and technology, from the traditional spontaneous development to a model featuring the integration of outside scientific and technical achievements into domestic R&D; from shutting the door to the outside world to opening the door; and from technonationalism to technoglobalism. This is the most important facet of China's transition from a marginalized country to one that is a strong power in the world of science and technology. China's opening to the Western capitalist countries is the main source of global technical innovation.[40] By directly obtaining advanced international technologies through multiple channels and leveraging the latecomer advantage to fully enjoy the major achievements of the world's science and technical revolutions, it is possible first to catch up with other countries technically and then do so economically.

Before 1980 China did not even have a patent system, and the number of invention patent applications submitted by Chinese citizens in 1980 accounted for only 0.7 percent of the world's total. As a developing country without the ability to technologically innovate, China mainly studied and imitated patented technologies imported from other countries, before integrating them into its own society.

Since opening the door to the outside world, however, China has secured more and more channels by which advanced international technologies can be obtained. At the same time, these channels have become more and more important. Among them, three stand out. One is foreign trade: importing hard technologies or importing technology-laden technical products and capital goods.[41] China has actively promoted trade liberalization and reduced its tariffs. On January 1, 2003, China introduced zero tariffs on imports of 213 categories of IT products. Two years later, on January 1, 2005, it scrapped tariffs on all imported IT products and lowered tariffs on most imported technical products. These actions stimulated imports of high technology. In 1990 China's advanced technology imports accounted for 2.0 percent of its GDP. By 2000 that proportion had risen to 4.47 percent, climbing to 9.45 percent by 2006.[42] China has also imported directly what is known as soft technology, such as international copyrights and licenses. In 1997 soft technology imports totaled $540 million, or 0.9 percent of the global aggregate. By 2004, however, soft technology imports had reached $4.5 billion, accounting for 3.7 percent of the world total.[43] Technology imports have indirectly stimulated domestic technical innovation and inspired direct participation in global technological competition. Over the course of the last fifteen years, the East Asia region has formed a production network, greatly raising the proportion of trade within the region and accelerating the diffusion of knowledge and technical innovation.[44] Dongguan City in Guangdong Province is held up by the World Bank as the most successful such case.[45]

Foreign direct investment (FDI) has a positive impact on the use of advanced foreign technologies.[46] According to UN Trade and Development Commission statistics, China's FDI inflow made up 0.5 percent of its GDP in 1980. Since then, the percentage has risen to 5.8 percent in 1990, 17.9 percent in 2000, and 19.7 percent (valued at $852.6 billion) in 2008.[47] The two kinds of technical capital in GDP—high-technology products import and FDI—rose greatly, from close to 2 percent in 1991 to nearly 8 percent in 2000 and 10 percent in 2008, before declining to 8.14 percent because of the international financial crisis (table 6-9). These numbers far exceed the proportion of GDP devoted to R&D expenditures (1.62 percent in 2009).[48] This indicates that China's technical capital mainly comes from foreign sources and FDI. The utilization of all kinds of technical sources in the world has significantly boosted China's S&T power.

Foreign patent applications in China constitute two-thirds of effective invention patents.[49] At the beginning of 2009, the number of foreign invention patent applications and the number of patents granted made up 27.2 percent and 49.1 percent, respectively, of all patents. By the end of 2009 the number of foreign invention patents in force reached 258,000, or 58.9 percent of the total.[50] This process encourages the transfer of patented foreign

Table 6-9. *Externally Acquired Technical Capital, China, Selected Years,*
1991–2009

Units as indicated

Source of capital	1991	1995	2000	2006	2008	2009
High-technology products import (billions of dollars)	n.a.	21.80	52.40	247.30	341.90	309.85
Share of GDP (percent)	2.00	3.00	4.47	9.45	7.90	6.31
Foreign direct investment (billions of dollars)	4.37	37.52	40.72	69.47	92.40	90.03
Share of GDP (percent)	0.96	5.15	3.40	2.64	2.14	1.83
Share of total in GDP (percent)	1.96	8.15	7.78	11.73	10.04	8.14

Source: *China Statistical Abstract 2010,* pp. 19, 65; Ministry of Science and Technology of the People's Republic of China, *S&T Statistics Data Book 2010.*

technologies to China to stimulate China's own technical innovation, allow-ing the PRC to avoid duplications and errors in R&D while simultaneously protecting and fully utilizing these foreign technologies. From 1991 to 2009 the number of foreign patent applications increased by a factor of 22.5, aver-aging 18.9 percent growth annually. The number of domestic patent applica-tions during the same period increased by a factor of 19.2 times, averaging 17.8 percent growth annually.[51] These statistics indicate that the transfer and diffusion of foreign patented technologies is favorable for stimulating domes-tic technical innovation. It is also a reflection of the great improvement in the protection of intellectual property rights in China. Patented technologies have become an increasingly important production factor and also a core element in boosting technical innovation capacity.

Despite the fact that China is not a world leader in the field of new tech-nologies and can even be considered a laggard, it is possible for it to realize a true leap in development as long as it opens its market, especially its technol-ogy market. For example, for the last twenty years China has lagged behind in the realm of information and communication technology (ICT), but with the opening of the domestic market and the importing of new technologies, China has succeeded in localizing these technologies and therefore has nar-rowed its gaps with developed countries and even overtaken the United States in the number of people using ICT products besides PCs.

With regard to knowledge innovation, China has adopted a strategy of international cooperation. According to the Science Citation Index, the num-ber of papers co-authored by Chinese and foreign scientists made up 20.1

Table 6-10. *Investment in Research and Development, by Sector, China, Selected Years, 2001–2006*
Percent

Year	Businesses	Large and medium-size enterprises	Small enterprises	Research institutions	Institutions of higher learning	Other
2001	63.3	60.5	2.8	20.8	17.5	1.2
2002	64.4	48.8	15.5	25.6	11.5	14.1
2003	68.4	63.0	5.4	18.9	12.6	5.4
2004	82.9	54.7	28.2	7.7	9.0	28.5
2005	74.4	61.2	13.2	16.8	8.6	13.4
2006	83.3	68.7	14.6	9.8	6.2	15.3

Source: Zhang Chunlin and others, *China: Encouraging Businesses to Make Innovations* (World Bank, 2009).

percent of all papers in 2008. The top six nations to cooperate with Chinese scholars in producing these papers were the United States (40.9 percent), Japan (12.0 percent), the United Kingdom (8.6 percent), Canada (7.8 percent), Germany (7.7 percent), and Australia (7.5 percent).[52]

THE BUSINESS SECTOR

With the rise of China's market economy and the decline of its planned economy, the business sector has replaced state-operated research institutions in dominating innovation activities.[53] Three developments in this area are particularly important. First, enterprises have become the main investors in R&D. The R&D investments of all types of enterprise have been rising steadily year by year (table 6-10). It stood at 60 percent in 2000 but rose to 71.7 percent by 2008.[54] Such investment has not only resulted in technical innovation but also enhanced the capacity of enterprises to obtain, use, and further develop external knowledge.[55] The main players in R&D are large and medium-size enterprises. Few small enterprises made any investments, although an increasing number are doing so. What prompted this increase in R&D investments by enterprises? A study by the World Bank finds that Chinese enterprises have a high rate of return on R&D investments. For every dollar invested, they get 0.70 dollar of total factor productivity. R&D investment also has a strong spillover effect.[56]

Second, enterprises have become the main players as regards R&D output. In 1995 the number of invention applications filed by industrial enterprises and mines made up 10.8 percent of the total. By 2000 the percentage had risen to 32.8 percent, climbing to 49.1 percent by 2008.[57] In 1995 technology transactions made up 0.44 percent of GDP. By 2008 these transactions comprised 0.89 percent of GDP.[58]

Third, enterprises have become the main group demanding research achievement. In 1995, 63.4 percent of the total transactions on the technology market were completed by enterprises. By 2009 that percentage had risen to 77.3 percent. In 2009, 86.6 percent of all contracted technology was exported by enterprises.[59]

GOVERNMENT GUIDANCE AND PROMOTION

In an increasingly competitive world, a country's science and technology power tends to mirror its national power. Most of the more than 200 countries in the world are market economies, but only a small number of developed countries with significant national power are truly strong science and technology powers. The big rise in China's S&T power is not only the result of fully utilizing market mechanisms; it also demonstrates that China, as a big developing country, has both clear national strategic goals and the means to realize them. These major advantages in China's development of S&T primarily manifest themselves in the following three ways.

First, the state has charted a course for long-term development and formulated a comprehensive national strategy.[60] In the context of more open and intense international competition, the Chinese central authorities have, in line with internal and external development trends as well as the demands of economic development and social progress, produced strategic goals for the step-by-step development of science and technology. Since 1978 the Chinese government has called for national science and technology conferences to set goals and to deploy resources to achieve these goals.

At the first national S&T conference in March 1978, Deng Xiaoping said that the modernization of S&T should be part of the four modernizations. He outlined the basic conditions for making China a strong modern power, further opening the doors to the outside world, and absorbing advanced foreign technology.[61] The second national S&T conference in 1985 made some important decisions regarding the across-the-board reform of the S&T system. It set a long-term policy of opening up to the outside world and securing a solid standing in the world.[62] The third national S&T conference in 1995 adopted "Decisions of the CCP Central Committee and the State Council on Accelerating the Pace of S&T Progress," which outlines the strategy of invigorating the nation by developing S&T. It delineates the strategic importance of S&T development and puts S&T development into high gear.

All of these decisions were aimed at enhancing national S&T power as well as improving the government's capacity to transform this power into economic productivity.[63] The national technical innovation conference held in 1999 and the paper "Decisions on Strengthening Technical Innovation and Developing High-Technology Industry," issued by the CCP Central Committee and the State Council, set stage-by-stage goals for medium- and long-term S&T development. The fourth national S&T conference in 2006 formulated a program that establishes the following:

—Technical progress as the first driving force behind economic and social development

—Independent innovation capacity as the central link in economic restructuring, the change of growth model, and the improvement of national competitiveness

—The building of an innovative country as the major future strategic objective

—The details of S&T development goals through 2020[64]

The fourth conference formulated a development strategy by analyzing new trends and new characteristics in world S&T development, setting specific stage-by-stage goals, and choosing a number of major areas and projects to address. In doing so, they charted the course for S&T development in China.

Second, the government has created a good climate for carrying out technical innovations by building a mechanism that effectively matches policies with incentives. This economic and social climate includes secure investment conditions, a stable macroeconomic order, a climate favorable for pioneering undertakings, and good public opinion on policy formulation. This confluence of factors helps to encourage the innovative and entrepreneurial spirit, inspire the spread and application of research achievements, and stimulate international cooperation and exchanges. The state has also formulated corresponding policies and measures to sustain such a development strategy. They include policies concerning financial support, tax concessions, bank loans, government procurement, foreign cooperation, training of personnel, and protection of intellectual property rights and necessary facilities. By providing incentives, the state has directed S&T forces toward realizing its strategic goals.

In 1999 the CCP Central Committee and the State Council issued decisions on strengthening technical innovation and developing high-technology industries. These decisions established a number of policies and incentives, including financial support, tax concessions, the management of scientific and technical personnel, evaluation and awards, and the management of intellectual property rights. In 2006, even before the central authorities made the decision to enhance independent innovative capacity by implementing the program for S&T development, the State Council made public specific policies concerning the implementation of the medium- and long-term S&T development program (2006–20). That December the central economic work conference called for increasing the protection of intellectual property rights and developing a large number of Chinese brands that would become famous across the globe. Two years later, in 2008, the State Council published a program for the protection of intellectual property rights, setting the following goals: making China the world leader in terms of patent grants by 2013; increasing the number of patent applications Chinese citizens applied for abroad; and establishing China as an advanced country with the ability to create, apply, protect, and manage intellectual property rights by 2020. The patent law that came into effect in October 2009 contains new specifications regarding the originality of patents in order to conform to world standards. At the same time, this patent law has eased the process through which foreigners can apply for patents in China. Countries that create intellectual property enforce it as well.[65] Not only has China become a prominent production base, it has also been transformed into a center of global innovation.

Third, the state provides financial support for R&D by attracting more nongovernmental funding sources. Not only has the state invested in major scientific and technical activities that directly support S&T development; more important, it has also invested in a way that stimulates input in S&T in other areas, such as localities. In the period between 1995 and 2009 (and especially 2000–09), the lever coefficient of the central investment in S&T (the ratio of local input in S&T to central input in the same area) grew significantly. In 1995 for every 1.00 yuan invested by the central government, localities invested 0.40 yuan. By 2005 the ratio rose to 1.00 to 0.65 and to 1.00 to 0.98 by 2008.[66]

Sustained High-Speed Economic Growth

High-speed economic growth includes both the huge demand for S&T generated by the demand side and the huge supply in S&T generated by the supply side; it reflects the complementary nature of economic and S&T power. In short, rapid accretions of economic power provide the economic base for improving S&T power, while the rise of S&T power stimulates economic growth. The expansion of these two powers, which are interdependent and mutually reinforcing, has turned China into a big economic power as well as a big S&T power.

R&D also has a scale effect in large countries; that is, it can efficiently utilize population size and market size for all public goods provided while developing labor-intensive, capital-intensive, technology-intensive, and knowledge-intensive (service) industries. There are three reasons for this: it can convert goods with a high fixed cost (such as R&D equipment) into goods with a low unit cost; it can raise returns on knowledge and technology investment via the large-scale diffusion of knowledge and the spirit of innovation; and it can generate positive externalities through the widespread use of technologies that yield high returns.[67] The best example of this is the hybrid rice and super-hybrid rice created by Chinese academician Yuan Longping. Today, areas in which hybrid rice is sown account for 50 percent of the total area where rice is sown in China, yielding 350 billion kilograms of rice.[68] According to studies by Fan Shenggen, 1.00 yuan of investment in agricultural R&D yields 5.59 yuan in added value, the biggest return on any investment of any kind.[69]

Among the world's developing countries, China is, indeed, an S&T success story. The aforementioned driving factors have enabled China to realize a great leap forward in S&T development, narrowing the gap between the PRC and the developed world.

China 2020: An Innovative Country

In 2006 the Chinese government made a major strategic decision to build an innovative country. That decision was based on the following assumptions: the first twenty years of this century represent a period of strategic opportunity for China and its S&T development; and the national development strategy—namely establishing independent, innovative capacity as the foundation of the development of S&T—will stimulate industrial restructuring, lead to a shift in the economic growth model, and help build a more resource-efficient and environment-friendly society.[70] This process requires turning factor-input-driven economic growth (especially resources and capital input) into total factor productivity, which mainly reflects technology progress or innovation-driven growth.

For this purpose, in 2006 the Chinese government formulated a program for medium- and long-term S&T development, with "building an innovative country by 2020" as its goal. In doing so, the Chinese government drew a clear road map to develop from a "big world S&T power" to a "strong world S&T power."[71] The ultimate goal is to enhance China's comprehensive national power and international competitiveness.

By 2020 China will realize the modernization of S&T and build the largest S&T contingent in the world, including personnel in all relevant disciplines.[72] Its science and technology workforce will grow from 60 million to about 120 million, with its proportion in the total employment rising from 8 percent to 13 percent. China will gradually increase its R&D investment, with its proportion of GDP first surpassing 2.0 percent, greater than the world average (1.6 percent), and then climbing above 2.5 percent by 2020, exceeding the average of OECD countries (2.2 percent).[73] It will bring its S&T power to over 60 percent and the number of annual Chinese patent grants and international citations into the top five in the world.[74] The GDP proportion of added value in high-technology industries will rise from 4.4 percent in 2008 to over 10 percent; the proportion of added value in the knowledge-intensive service industry will rise from 10.6 percent in 2007 to over 15 percent. Export of high-technology industries with independent intellectual property rights and high value-added ratios will increase, and China will become one of the largest and the most competitive countries for high-tech production, manufacturing, design, R&D, and exporting.[75]

At the same time, China will accelerate the pace of spreading, converting, trading, popularizing, and applying new research achievements, expanding trading in domestic technology markets, and bringing the proportion of transactions as a share of GDP from 0.85 percent to over 2.5 percent. By 2020 its information industry will be modernized, the first industry to reach such a state. As a result, China will be the largest society featuring the widespread application of technology. It will also be the largest information society, with a network that covers the entire country. Finally, it will be an innovative knowledge society, with its people absorbing and using innovations so as to have the knowledge capital necessary for being a moderately developed country by 2050.

The long-term goal for China's S&T development is to make China the world's largest innovative society and knowledge society. Such a society would make full use of global knowledge and technical resources, mobilize all of its technical forces, and intensify the input of knowledge, technology, information, and infrastructure. This would increase innovative power, expand the scale effect of innovation and knowledge, raise the knowledge capital of the people, and quicken the pace of catching up with developed countries in the area of knowledge.

The PRC will strengthen enterprises as the main players in innovation and encourage them to enter into cooperation with research institutions and universities. It will make greater efforts to develop its intellectual property system, improve its technical standards system, launch major innovation projects, and develop a system to provide support for S&T development, such as a government-funded incubator and a product transaction market. China will increase the dissemination of knowledge and technology, especially among the rural population, raising the scientific and cultural qualities of all Chinese citizens. It will commit enough resources to cultivate internationally top-rated technical personnel and world-class scientists and leaders in S&T. It will actively try to attract highly trained foreign intellectuals and the nearly million Chinese students studying abroad.[76]

The government will provide support, especially financial support, to research in basic sciences, frontier technologies, and social welfare. It will promote the humanities, philosophy, and social sciences and encourage and ensure the implementation of the policy known as "Letting a hundred flowers blossom and a hundred schools of thought contend." In doing so, it will encourage and promote theoretical innovations and the integration of the natural sciences with philosophy and the social sciences. As innovations and knowledge have a spillover effect, and as China enjoys favorable economies of scale, the innovation and knowledge of one person, one enterprise, and one organization may benefit the whole society. As a result, the encouragement of innovation may enhance private gains and enable everyone to benefit from them.

The future of the world lies with China.[77] The future of China lies in innovation. And innovation depends on knowledge contributions. China's most important goal in the twenty-first century is to make contributions to human knowledge, like its four major inventions of yesteryear—the compass, gunpowder, papermaking, and printing. Climate change is one of the greatest threats ever faced by society. As human technology caused the problem, so too can human technology find the solution.[78] China, with the world's largest human resources in S&T, should work with the rest of the world to launch a new green revolution and take the lead in carrying out green innovations to address global climate change.

CHAPTER SEVEN

CLIMATE CHANGE AND SUSTAINABLE GROWTH

G lobal climate change is the largest constraint on and challenge to the future of Chinese economic and social development. The rise of China has already changed the international order. Such a seismic shift in global power has brought about two grave challenges, which will have to be faced by both China and the world at large. The first lies in China's need to ensure energy security and protect its environment. The second is the global need to combat climate change. Against the backdrop of an increasing number of voices calling for action to mitigate climate change and the active promotion of these voices by certain big powers, Chinese scholars are still basing their studies of the issue on the principle of common but differentiated responsibilities.

China, as a latecomer to industrialization, has been afforded the chance to avoid the mistakes made by early industrialized countries. In addition, it has the opportunity to create a new development model: a low-carbon economy. In seeking its own green development and green "rise," it can greatly contribute to the worldwide green revolution. If China fails in this endeavor, the world will fail to combat climate change. The best way to avoid failure is not to deemphasize development but rather to effectuate a fundamental change in the model of development—a change from "black" development to "green" development, a change from depleting nature to harmonious coexistence with nature, and a change from ecological deficit to ecological surplus.

All of this can be achieved by following a scientific approach to development. In the coming years, China is likely to become both the world's largest energy consumer and the world's largest investor in traditional energy and renewable energy development, overtaking the United States. At the same time, the PRC is and will remain for some time the biggest greenhouse gas (GHG) emitter. China's rise has both positive and negative externalities. The positive externalities include China's contributions to global economic, trade, and investment growth; the negative side is the basic economic principle that everything has a cost.

This chapter discusses the climate challenge China faces in its process of modernization and how it should meet that challenge. It also discusses the path to green development and how China will contribute to the international green revolution.

Climate Change: The Biggest Global Challenge

From a long-term perspective, the global ecological crisis, especially climate change, is the largest, most formidable, and most fundamental challenge to the continued existence of mankind.[1] In contrast, the financial crisis is short-lived and less stark. Nevertheless, the financial crisis is likely to overshadow climate change for some time, impacting the ability of many countries, especially developing countries, to combat climate change.

In 2007 the Intergovernmental Panel on Climate Change (IPCC) published its fourth climate change assessment report. Written by the world's top scientists, who were brought together by the IPCC based on their cooperative studies, the report arrived at several scientific conclusions.[2] The IPCC report has made us all aware that human activities have caused global climate change. Seeing as GHGs are already in the air and it is impossible to dispose of them in a short time, no matter what measures are taken, global warming can only get worse. The challenge will be even graver during the Twelfth Five-Year Plan.[3] According to the IPCC report, which compares the years 1980 and 1999, the world's average surface temperature may rise by 1.1–6.4 degrees centigrade and its sea level by 18–59 centimeters by the end of the twenty-first century.

According to the report, China will be one of the countries most impacted by global climate change. It has already shrunk the glaciers of the Himalayas, and rising temperatures have shifted the temperate and arid zones northward. Cities such as Shanghai will experience more frequent and more serious heat waves, making the rapidly expanding urban population uncomfortable. A UNDP report goes even further, pointing out that, based on the current speed of melting, two-thirds of the glaciers, including those on Tianshan, will disappear by 2056, and all of them will be gone by the year 2100. The glaciers on the Qinghai-Tibet plateau are barometers of global climate change and also the sources of the Yellow and Yangtze rivers. They are currently shrinking at an annual rate of 7 percent. In any climate change scenario, if the temperature rises by more than 2 degrees centigrade, the danger threshold, the glaciers will melt even more quickly. With the exhaustion of this water bank, the water flow emanating from it would be reduced. The seven major waterways in Asia that supply water and make possible agricultural growth for more than 2 billion people—Yarlung Zangbo, Ganga (Ganges), Salween, Yellow, Indus, Mekong, and Yangtze—will all be affected. The report predicts that by 2050 the flow of Yarlung Zangbo will be reduced by 14–20 percent.[4]

Arid and semiarid areas, which make up half of the land territory of China, will be affected directly. Precipitation over Japan's 378,000 square kilometers of land averages 1,000 millimeters annually. Semiarid to semihumid climate regions of subtropical and tropical latitudes tend to have rainfall between 750 millimeters and 1,270 millimeters a year.[5] But in China humid areas make up only one-third of its territory. Only coastal areas in southern China have a mean annual precipitation of 1,600–2,000 millimeters. The Yangtze River basin and the areas immediately south of it average 1,000 millimeters of precipitation. Northern and northeastern China average 400–800 millimeters; northwestern China, 100–200 millimeters; and the Tarim, Turpan, and Qaidam basins, less than 25 millimeters.[6] Given similar temperatures, China will be hit worse than Japan.

Northern and western China will be seriously affected by climate change. The aforementioned UNDP report points out that northern China suffers greatly from water shortages. In parts of the Hai River, Huai River, and Yellow River basins current water demand is 140 percent of the available recycled water supply, which accounts for the rapid dwindling of the water sources of the main waterways and a significant drop in the subterranean water table. From a midterm perspective, the shrinking of glaciers will make the water shortages even worse. About half of China's 128 million rural poor people live in this area, an area whose cropland accounts for about 40 percent of the national total and whose output value makes up one-third of national GDP.

The shrinking of glaciers will have a serious impact on human development. All of the ecosystems in western China will be threatened. Experts predict that by 2050 the temperature of this area will rise by 1.0–2.5 degrees centigrade.[7] The Qinghai-Tibet plateau is as big as the whole of Western Europe and boasts more than 45,000 glaciers, which are disappearing at the alarming rate of 131.4 square kilometers a year. China's national plan for coping with climate change also points out that, over the past half century, northwestern Chinese glaciers have shrunk by 21 percent and the permafrost in Tibet has become 4–5 meters thinner at its maximum.[8]

Eastern China, represented by Shanghai, is especially threatened by climate change–related disasters. Shanghai is situated at the mouth of the Yangtze River, sitting only four meters above sea level. It is an easy target for summer typhoons, storm surges, and floods. Indeed, the 18 million people in the city are continually threatened by floods. In addition, rising sea levels have led to increasing storm surges. Shanghai has been listed as among the most vulnerable cities in the world to the effects of climate change.[9] The most vulnerable residents of the city are the approximately 3 million migrant workers temporarily living in Shanghai. Most of them live in makeshift dwellings near construction sites or in previously flooded areas. Their rights are limited, affecting their ability to resist disasters. These migrants will be subject to the greatest danger if disaster strikes.

China: Worst Victim of Climate Change

China's national plan for combating climate change categorizes the basic climate-related national conditions as poor climatic conditions, serious natural disasters, vulnerable ecosystems, a coal-dominated energy structure, and a large population with low economic development levels.[10] With the world's largest population, a vast territory, and natural ecosystems, China is one of the countries most vulnerable to global climate change. China stands out in terms of its frequency of meteorological disasters caused by rain, drought, snow, extreme heat or cold, ice, or wind that can result in blizzards, cyclonic storms, droughts, hailstorms, heat waves, and tornadoes. Climate change affects China's agriculture, livestock breeding, forestry, water resources, coastal areas, and ecosystems. Although climate change has had some positive effects, such as extending the growing period for some crops and shortening the frost and freezing periods, on the whole it produces much more harm than benefit.

NATURAL DISASTERS SINCE ANCIENT TIMES

The history of China is a history of natural disasters. During the more than 2000 years between the Zhou dynasty and the Qing dynasty, China was hit by 1,052 droughts, 1,029 floods, and 473 locust plagues. Generally, these disasters increased in frequency over time, though there have been big fluctuations. Analyses of disasters vis-à-vis population show that the population figure drops during periods of severe disasters. When population increases the reclamation of land from lakes, rivers, and the sea also increases, damaging the delicate ecological balance of nature.[11]

METEOROLOGICAL DISASTERS

Oxfam recently published a study reporting that between 1997 and 2008 an average of 278 million people a year were affected by climate-related disasters.[12] By 2015 that number is expected to rise by 45 percent to 375 million, which will undoubtedly pose major challenges to the global humanitarian relief system.

Between 1990 and 2007 an average of 120 million people a year were affected by climate-related natural disasters in China, 52.4 percent of the total number of individuals impacted globally. Though this figure fluctuates from year to year, it has generally assumed an upward trend. This indicates that China is prone to increasingly large climate-related disasters and that its proportion of people affected by such disasters will likely continue to rise (table 7-1).

Climate-related disasters affect the largest areas relative to any other natural catastrophe. Of the top ten natural world disasters between 1990 and 2008, in terms of number of people impacted, the most frequent were floods and droughts, and these were mostly concentrated in China and India (60 percent occurred in China).[13] After 1950 the frequency of major disasters in China has steadily trended upward. These natural disasters took place in the 1980s, the 1990s, and at the beginning of the twenty-first century. This shows, to a certain extent, the abnormal changes to climate after the advent of the twenty-first century.

NATURAL DISASTERS AND AGRICULTURAL LOSSES

After the founding of the People's Republic of China the number of areas affected by floods and droughts increased through to the present. I use the term *disaster area* to refer to regions where natural disasters caused at least a 30 percent reduction in food production when compared with a normal year. Thus not all areas affected by natural disasters are disaster areas. In the 1950s close to 8 million hectares of land a year were struck by floods. That figure climbed to 12 million hectares in the 1990s and at present stands at 10 million hectares. At the same time, the areas affected by droughts have also steadily increased, reaching more than 23 million hectares at present.

The average annual reduction in food production can be calculated according to disaster areas. Table 7-2 shows that in the 1950s natural disasters caused a loss of nearly 4 million tons of grain output, equivalent to 2.1 percent of the decade's total grain output (standard error = 0.8). Yet over the past five years

Table 7-1. *Population Affected by Climate-Related Disasters,
China and the World, 1990–2007*[a]
Units as indicated

Year	China population (millions)	World population (millions)	China's share of world population (percent)
1990	47.69	81.93	58.2
1991	214.34	281.34	76.2
1992	28.38	66.76	42.5
1993	8.50	170.99	5.0
1994	220.04	266.32	82.6
1995	147.50	220.04	67.0
1996	165.30	213.56	77.4
1997	14.29	68.71	20.8
1998	225.24	339.15	66.4
1999	125.92	269.08	46.8
2000	23.17	170.55	13.6
2001	39.25	98.91	39.7
2002	285.18	659.20	43.3
2003	214.78	250.77	85.6
2004	51.93	158.52	32.8
2005	83.03	153.72	54.0
2006	87.83	118.07	74.4
2007	120.04	209.96	57.2
Average	116.80	210.98	52.4

Source: Shamanthy Ganeshan and Wayne Diamond, "Forecasting the Numbers of People Affected Annually by Natural Disasters up to 2015," 2009 (www.oxfam.org/sites/www.oxfam.org/files/forecasting-disasters-2015.pdf).

a. Climate-related disasters include drought, extreme temperature, natural disasters, floods, windstorms, and displacement of people resulting from wet weather. Natural disasters covered by the statistics satisfy at least one of the following conditions: at least ten deaths are caused, at least a hundred people are affected, an emergency has been called, international aid is asked for.

grain output has rapidly declined. Now, due to the steady rise in unit output, a natural disaster of a magnitude similar to that of the 1950s may cause greater losses. So in the face of accelerated natural disasters and rising land productivity, disaster prevention and mitigation are necessary to achieve an increase in total grain output. China is now the biggest cereal and food producer in

Table 7-2. *Average Annual Grain Reductions Due to Natural Disasters, China, Selected Periods, 1952–2006*[a]
Units as indicated

Period	Annual grain reduction (million tons)	Total annual grain output (million tons)	Loss (percent)[b]
1952–59	3.79	180.25	2.1
	(1.52)	(14.37)	(0.80)
1960–69	6.12	173.86	3.5
	(1.71)	(25.97)	(1.0)
1970–79	6.63	276.12	2.4
	(3.56)	(28.97)	(1.3)
1980–89	15.95	276.99	4.2
	(3.25)	(32.54)	(0.9)
1990–2000	32.91	470.36	(7.00)
	(6.41)	(29.03)	(1.37)
2001–06	34.04	465.23	7.38
	(7.57)	(23.78)	(1.91)

Source: National Bureau of Statistics of China, *China Statistical Yearbook,* various years; author's calculations.
a. Standard errors are shown in parentheses.
b. Grain loss is calculated at 30 percent of the normal year's output.

the world. It is also the country with the biggest losses resulting from natural disasters. The most serious threats to China's agriculture and the livelihood of its farmers are natural disasters and other forms of abnormal weather caused by global climate change. The study of disaster prevention and mitigation is, therefore, of great import.

Natural Disasters and Economic Loss

During the 1990s direct economic losses caused by natural disasters grew from 67 billion yuan to more than 300 billion yuan (table 7-3). At the same time, however, economic growth led to a steady drop in direct economic losses as a proportion of GDP, from 3–4 percent in the 1990s to about 1 percent in the 2000s. Direct economic losses caused by natural disasters in 2005 were 204.21 billion yuan, or 1.11 percent of GDP. In 2006 they were 252.81 billion yuan, or 1.21 percent of GDP, while in 2009 they were 252.4 billion yuan, or 0.74 percent of GDP. The proportion of direct economic losses

Table 7-3. *Direct Economic Losses Caused by Natural Disasters and their Proportion of GDP, Selected Years, 1990–2008*
Units as indicated

Year	Economic loss (billion yuan)	Proportion in GDP of the year (percent)	Proportion in added GDP of the year (percent)
1990	66.6	3.6	40.6
1991 (flood)	121.5	5.6	39.6
1992	85.4	3.2	17.0
1993	99.3	2.9	12.4
1994	187.6	4.0	15.5
1998 (flood)	300.7	3.8	62.9
2003 (SARS)	148.2	1.1	10.9
2005	204.2	1.1	10.9
2006	252.8	1.2	10.9
2007	236.2	0.96	8.1
2008	1,175.2	3.9	47.3

Source: National Bureau of Statistics of China, *Report of the Damage Caused by Disaster in China: 1949–1995* (1995), pp. 403–07; National Bureau of Statistics of China, *Statistical Bulletin on Economic and Social Development;* author's calculations.

caused by natural disasters in GDP has dropped steadily. The year 2008 saw the biggest rise in direct economic losses caused by natural disasters, totaling 1.175 trillion yuan, including 845.1 billion yuan caused by the Wenchuan earthquake, which registered an 8.0 on the Richter scale.[14]

POOR PEOPLE: THE MOST VULNERABLE TO CLIMATE CHANGE

Most poor people in China live in mountainous areas, loess highlands, or remote barren areas where the natural conditions are harsh. In all of these geographic regions, ecology has been seriously damaged and land productivity is low. In addition, areas hit frequently by natural disasters are at high risk for the spreading of endemic diseases. In 2005 dry land made up 67 percent of the area in these areas, and sloping cropland with an incline of twenty-five degrees or more made up 15.19 percent of total arable land; 24.16 percent of the villages did not have irrigated land; 34.13 percent of the households did not have rice paddies or irrigated cropland; 53.14 percent of the villages were hit by natural disasters; and 14.19 percent of the villages lost at least half of their agricultural output. Among poor households, 41.11 percent were struck by natural disasters.[15]

Even with the decrease in the rural poor population, those who are still in poverty live mostly in areas with harsh natural conditions. Such poverty-stricken areas are associated with vulnerable ecological niches. These niches are on the border of ecological zones, where ecological stability is poor, various factors interact with one another violently, resistance to external interference is weak, and the biosystem's ability to bear human activity is low. In addition, poor management of resources in the economic development of these areas has caused the deterioration of their ecology and environment. Thus China's poor areas are more sensitive to the impacts of climate change. Extreme weather events, including droughts and floods, occur with greater frequency and intensity in these areas. This not only poses a long-term challenge to China as it seeks to adapt to climate change and improve its ability to resist natural disasters but also makes poverty relief difficult.

China: A Superpower in Greenhouse Gas Emissions

China is the largest coal consumer and the largest emitter of SO_2 in the world.[16] Calculations by the International Energy Agency (IEA), based on preliminary data, show that China had by 2010 overtaken the United States to become the world's largest energy user as well as its biggest carbon emitter.[17] The Chinese government has dismissed IEA's estimates, undoubtedly making China the target of international castigation.[18] According to the IEA, China's energy consumption growth between 2000 and 2008 was 34 percent of global growth. At the same time, the percentage of coal consumption was over 68 percent of the world total; and 52 percent of new global CO_2 emissions come from China.[19] China is the world's largest "black cat" (referring to coal consumption) and has the biggest negative external effect in the world.

With the deepening of globalization, the international division of labor has brought greater pressure to bear upon China's resources and environment. Under the new pattern of near complete economic integration and specialized division of labor, there has not only been a shift of multinational industries to Asia but also a higher rate of energy consumption and emissions in the PRC. According to the *World Energy Outlook 2007: China and India Insights,* China's energy reexport was 400 million tons of oil equivalent (Mtoe), about 25 percent of the year's total energy consumption. The energy contained in China's imports was 171 Mtoe, about 10 percent of the year's energy demand. The energy contained in China's exports is far higher than other countries (in 2001, energy used by the United States was 6 percent of the world total; EU, 7 percent; Japan, 10 percent). The higher proportion of energy reexport also resulted in higher CO_2 emissions. The CO_2 emissions in China's energy reexport were 430 million tons, 26 percent of the carbon intensity produced by all sectors in 2001.

A study published by the Tyndall Centre for Climate Change Research shows that in 2004 about 1.11 gross tonnage of CO_2 emissions could be attributed to China's net exports, which accounted for 23 percent of the year's CO_2 emissions (4.73 gross tonnage).[20] The carbon emissions resulting from China's net exports made up a quarter of the year's total. This figure is equal to the total emissions of Japan in the same year, the combined totals of Germany and Australia, and two times that of Great Britain. Li Liping, Ren Yong, and Tian Chunxiu of the Ministry of Environmental Protection point out that the above study covers only direct emissions, neglecting the emissions created during the production process, which were likely significant.[21] Carbon emissions from trade projects, the biggest contributor to exports, are slightly lower than the average carbon emissions in terms of intensity. This figure, therefore, has to be verified by a more complete input-output model. Even so, these two figures should not differ too significantly.

The emissions resulting from China's energy reexport mean that a considerable part of energy consumption and carbon emissions that originally emanated from European, American, and Japanese manufacturing has been transferred to China. This reflects the fact that the regional structure of global energy consumption and GHG emissions is shifting (table 7-4). In other words, the proportion of emissions coming from Europe, the United States, and Japan is dropping while that of China, India, and other big developing countries is rising.

China's Policies Combating Climate Change

As the climate change challenge grows, China's climate policy is also undergoing great changes. These changes include shifting from a meteorological forecast and disaster relief policy to a climate change policy; moving from an energy conservation policy to an integrated energy efficiency and emissions reduction policy; and including policies regarding energy, environmental protection, and meteorological sectors in the national development strategy.

The emission reduction targets set by China are very difficult to achieve. The Eleventh Five-Year Plan set the target of reducing unit GDP energy consumption by 20 percent, from 1.22 tons of standard coal down to 0.97 ton.

Table 7-4. *CO$_2$ Emissions, Six Major Economies, Selected Years, 1960–2030*[a]
Percent of world total

Country	1960	1970	1980	1990	2005	2015	2030
China	8.98	5.65	8.08	11.29	19.16	25.34	27.32
European Union	15.87	15.09	13.59	10.96	14.82	11.77	9.97
United States	33.68	31.18	25.32	22.67	21.75	19.76	16.44
Japan	2.47	4.96	4.71	4.76	4.55	3.79	2.82
Russia	n.a.	n.a.	n.a.	9.26[b]	5.74	5.28	4.71
India	1.28	1.30	1.79	3.01	4.14	5.28	7.88
Total	n.a.	n.a.	n.a.	n.a.	70.16	71.23	69.14
EU/China gap	1.77	2.67	1.68	0.97	0.77	0.46	0.36
U.S./China gap	3.75	5.52	3.13	2.01	1.14	0.78	0.60

Source: World Bank, *World Development Indicators 2006;* International Energy Agency, *World Energy Outlook 2007.*
a. The reference scenario (according to the current state, without relevant policy for controlling emissions) takes the European Union as including twenty-five countries.
b. 1992 figures.

This requires that the growth of total energy consumption be kept at or below 5.2 percent even if GDP growth averages 10 percent annually. In contrast, the EU has only committed to reduce per capita energy consumption by 20 percent before 2020.[22]

The steady drop in per capita GDP energy consumption shows that China's energy utilization rate has been rising since reform and opening. Per capita GDP energy consumption dropped by 23.5 percent, 11.9 percent, 25.5 percent, and 34.3 percent during the Sixth, Seventh, Eighth, and Ninth Five-Year Plans, respectively. But the trend was reversed during the Tenth Five-Year Plan, when it rose by 8.5 percent due to the abolishment of an indicator on energy conservation. It is a heavy historical lesson.

Obligatory targets are of paramount importance for restricting government. One of the important innovations of the Eleventh Five-Year Plan is the definition of government responsibility targets as obligatory. The most important of these targets include a 20 percent reduction of per unit energy consumption and a 10 percent reduction in emissions and discharge of pollutants. The plan states: "The obligatory targets set by the program are legally binding and should be made part of the criteria for the integrated evaluation and performance assessment of the economic and social development of all regions and departments." In reality, the 20 percent reduction in energy

Table 7-5. *Renewable Energy Consumption, China, Selected Years, 1990–2008*
Units as indicated

Item	1990	2000	2003	2008
Renewable energy consumption (10,000 tons of standard coal)	6,350	11,504	18,344	25,059
Renewable energy in total energy consumption (percent)	6.14	8.14	10.8	8.8
U.S. renewable energy consumption in total primary energy supply (percent)	5.1	4.4	4.2	n.a.

Source: *China Energy Development Report 2007; China Energy Development Report 2009*, p. 26; OECD/IEA, *Renewable Information 2005.*
n.a. = Not available.

consumption during the Eleventh Five-Year Plan is not only an economic target but also a political commitment by the Chinese government to both its own citizens and the people of the world. It demonstrates the political will and determination of the Chinese government to promote energy conservation and emissions reduction, cope with climate change, and develop a low-carbon economy. This target is the first bold attempt to create a low-carbon economy. Its political significance far overshadows its economic significance.

China's energy structure is dominated by coal, with only a fraction consisting of nonfossil energy and clean energy (table 7-5). In 2008 the percentage of nuclear power was less than 2 percent, with natural gas standing at 3.8 percent.[23] Both percentages are far lower than those found in developed countries. China is facing growing pressure to reduce GHG emissions. Thus it has accelerated the development of nuclear power and is doing its utmost to develop wind power, solar power, and biomass power. Its goal is to increase the percentage of nonfossil fuels in the energy structure to 15 percent by 2020 and 30 percent by 2030, making it the main energy source in the national energy structure. The policy should be to reduce the proportion of coal and thermal power, to raise the proportion of clean energy, to import quality energy, and to realize energy conservation in all areas. It should also tighten national energy consumption standards and market access criteria by reducing the energy consumption per unit of manufactured products. Finally, it should seek to reduce net energy exports, which stood at 18 percent in 2001 and 28 percent in 2004, and their corresponding carbon emissions (about 1.1 billion tons, or 23 percent total global emissions).[24]

Forests are the biggest carbon-storage areas and the most economical carbon sinks. According to the computations of scientists, every cubic meter of woods can absorb 1.38 tons of CO_2 and discharge 1.62 tons of oxygen. Experts at Beijing University state that the cost of one ton of CO_2 stored in a hectare of forest is only 122 yuan.[25] It is both highly efficient and cheap to sink carbon in forests. Forests are also good for creating jobs, and President Hu Jintao has called for a forestry and climate cooperation network in the Asia-Pacific region.

The rapid growth of forests has greatly raised China's carbon-sinking capacity and potential for reducing emissions. Data from China's State Forestry Administration show that from 1999 to 2005 China was the fastest-growing country in terms of forest resources. These forests absorbed a large amount of CO_2 and proved to be of huge ecological value to the sustainable social and economic development of both China and the world at large. China's forest carbon-sinking capacity increased from 136.42 tons per hectare at the beginning of the 1980s to 150.47 tons per hectare at the beginning of the twenty-first century. The trees planted in the 1980–2005 reforestation drive are estimated to have absorbed about 4.68 billion tons of CO_2. Indeed, preventing the destruction of forests has allowed China to reduce CO_2 emissions by 430 million tons. In 2004 the forests in China absorbed about 500 million tons of CO_2, more than 8 percent of the total GHG emissions that year.[26] From 1977 to 2003 China's standing stock of timber reached 40 percent, which means that the carbon-storing capacity of the forests also increased by 40 percent.

China's Roadmap for Tackling Climate Change

Global climate change is growing increasingly dangerous. Its impact far overshadows the negative effects of the global financial crisis. Seeing it as the biggest constraint to the long-term development of the PRC, the Chinese government found it necessary to formulate long-range and strategically forward-looking climate change policies and targets, drawing a clearer roadmap for reducing GHGs.

The general line is, "One world, one dream, one action," the action referring to emissions reduction. China must leave behind its narrow view of emissions reduction and instead place it within the historical context of the country's modernization.

BREAKING THE STALEMATE AND WORKING TOGETHER
TO REDUCE EMISSIONS

Accelerated global warming poses a huge challenge to the world. I believe that if international cooperation is to succeed, it is necessary to divide countries into four groups according to the Human Development Index (HDI), which will make more than 60 percent of countries and populations implement emissions reduction either unconditionally or conditionally. Developed countries, especially those in the high-HDI group, should reduce emissions unconditionally. Countries with an upper middle level of HDI should reduce emissions conditionally. The majority of the largest twenty emitters are in the first high-HDI group. China and India are also on the top-twenty list, but they are in the second and third HDI groups, respectively. No matter what kind of standard is adopted, it is time for the whole world to reduce emissions.

The world needs to take advantage of all windows of opportunity to address climate change. What position should China take? I think that China should be intimately involved in the international negotiation, with its decisions guided by the basic idea that we are all in the same boat. Such involvement requires that China give full consideration to a range of climate policy options in mapping out the Twelfth Five-Year Plan, including clearer emissions reduction targets for various regions and the country as a whole so as to advance China's green development and involvement in the world's green development, the mapping of an emissions schedule (for 2020, 2030, and 2050), and the translation of challenges into opportunities and of crises into turning points.

It is necessary to translate China's dream into the world's dream. When we say "one world," we mean that the earth is growing smaller and smaller. "One dream" refers to a green world, and "one action" refers to emissions reduction. Since there was no agreement on the reduction of carbon emissions in Copenhagen, we must look forward to and prepare for the upcoming Mexico City summit. Mankind needs to achieve a shared understanding and set of shared principles so that we can take concerted action against that which threatens both human and ecological security, namely climate change.

By 2020 China's per capita GDP may reach the world average and its HDI will reach 0.87–0.88, putting the PRC in the high-level HDI group. This reflects the most important feature of China's development model: living standards have reached a high level despite per capita income levels that still lag behind those of developed countries. At the same time, intracountry regional HDI will rise to varying degrees, with about 70 percent of the population to be listed in the first group, the population located in the second group dropping from 70 percent to 20 percent, and the population occupying the third group disappearing. If everything goes well, places such as Guizhou and Tibet, which are in the lowest level of HDI in the world, are likely to edge into the second group or approach it. Viewed from an international perspective, it is entirely possible for China to realize its goal of building a well-off society (in every sense of the word) by 2020. Looked at from the perspective of emissions reduction, China's betterment will enhance its ability to meet its global obligations.

China is facing a role paradox with regard to global emissions reduction. On the one hand, China is the largest GHG emitter in the world and thus has a heavy responsibility for emissions reduction. Without China's participation and action, developed countries cannot succeed in their actions, and any plan to reduce emissions will end in failure. On the other hand, China is a developing country. It wants neither to take the lead in lowering emissions nor offend other developing countries (especially India), which would dissuade them from making open commitments to emissions reduction. Instead, China makes strong demands on developed countries to reduce emissions by 40 percent. China is already a pollution "superpower," the number-one polluter in the world as of 2006.[27] Its CO_2 emissions have grown since 1990 faster than those of any other country.[28] By 2030 China's GHG emissions could reach 12.8 billion tons of CO_2, accounting for 30.5 percent of the world total. This is the biggest negative externality resulting from the rise of China and also the biggest threat brought about by future global climate change. China's GHG emissions represent the largest negative factor undermining its international soft power. It is therefore imperative for China to be self-disciplined and self-constrained instead of developing at will. There should be no laissez-faire development.

Some Chinese scholars hold that global climate change is a problem of "theirs" (developed countries) and not "ours" (developing countries). My view is that the global climate change problem is both theirs and ours. It is an especially severe problem for China, which is most vulnerable to natural disasters and air pollution. The more than one billion people who live in China are the worst victims of "our" problem. Developed countries should be compelled to reduce emissions, and developing countries should be encouraged to reduce emissions under a multilateral agreement. This represents both the biggest external pressure and challenge and the best opportunity for China. Working together with the rest of the world will make China an initiator, a leader, and an innovator in global actions to reduce emissions.

The United States has already taken the lead in emissions reduction. The world now needs China to take the lead.[29] For this purpose, China should work together with the five major GHG-emitting countries to cut GHG emissions, even if China's per capita emissions are far lower than those of developed countries.

CLIMATE CHANGE POLICY OBJECTIVES FOR 2020

China should formulate a climate change policy framework according to its core interests and national conditions. At the same time, China should, in the interest of human development, not only join the rest of world but also try to lead the rest of the world in green development. China needs to study and expand its National Scheme for Tackling Climate Change (2011–20), which was issued by the National Development and Reform Commission in 2007 and includes the following five objectives:

—Energy conservation: reduce per capita GDP energy consumption by 20 percent every five years, with the 2006–20 cumulative reduction equaling 50–80 percent.

—Emissions reduction: reduce discharges of major pollutants by 10 percent every five years, with the 2006–20 cumulative reduction of SO_2 equaling 30–40 percent and of CO_2 equaling 50 percent.

—Innovative green technologies: become a collaborator, a leader, and a user.

—Green energy market: become the world's largest wind power and solar power market, become a producer and exporter of new energy technologies and equipment, and increase the use of clean energy to 20 percent of all energy consumed.

—Green ecology: build the world's biggest artificial forest carbon sink and the world's largest green screens (shelter belts in the northeast, the north, the northwest, and the southeast coast).

OTHER POLICIES ASSOCIATED WITH CLIMATE CHANGE

In addition to the above objectives, China should also pursue the following eleven policy initiatives:

First, the Chinese government should focus on reforming the industry structure to lower the proportion of industrial added value in GDP. Currently, China's energy consumption per unit of industrial added value is one of the highest in the world. For instance, in 2008 industrial production made up 42.9 percent of GDP, while industry's energy consumption accounted for 74 percent of the national aggregate.[30] In other words, industrial energy consumption was 1.67 times that of per unit GDP, or 4.91 times the per unit added value of the service industry.[31] Pollutant discharge per unit of industrial added value was also one of the largest. The wastewater discharged by industry in 2006 made up 44.7 percent of the national total; industrial chemical oxygen demands made up 37.9 percent of the national total; industrial emissions of ammonium nitrogen accounted for 30 percent of the national total; industrial emissions of SO_2 made up 86.3 percent of the national total; and industrial soot emissions accounted for 79.4 percent of the national total.[32] China should focus on strictly limiting or eliminating the following industries: iron and steel, building materials, nonmetallic ores, chemicals, and petrochemicals, in order to reduce industrial energy consumption, carbon emissions, and the use of capital.[33] China will instead focus on employment-intensive, high-technology, and new technology industries and will develop modern service industries, especially those that are information, knowledge, and employment intensive.

Second, the Chinese government should adjust its energy policy to raise the proportion of quality and renewable energy and reduce the proportion of energy with high carbon content. China is the dirtiest country in terms of its energy use. Its energy consumption structure is irrational, featuring dirty coal as the main form of energy. Such a structure entails huge negative social externalities and costs.[34] From 1996 to 2001 the government took the first steps in cleaning up this dirty industry by taking advantage of a drop in energy demand. As a result there was a steady drop in the percentage of coal in total energy consumption. After 2001, however, demand surged upward. It is therefore necessary to clarify the essence of future energy and industrial policies so that they limit the development of the coal industry, reduce coal consumption, and make coal consumption a compulsory indicator.[35] In addition, the state will raise the clean coal utilization rate.[36]

Third, China should improve its energy efficiency. This is necessary in order to realize the 2010 and 2020 goals for energy conservation. It can be accomplished through two strict policy interventions. One would make high energy–consuming industries reduce their energy consumption per unit of product obligatorily and the other would require that such standards be the necessary conditions for market access.

Fourth, the state should implement energy price reform policies. This will include the deregulation of coal and electricity prices and the establishment of government-set prices for natural monopoly industries such as power distribution. In addition, petroleum and natural gas prices should move closer to international market rates, and the mechanism used to set these prices should be improved. Finally, the wholesale and retail prices of petroleum should be deregulated in order to encourage market competition.

Fifth, the state should levy taxes on pollution. This would include establishing taxes on carbon and sulfur emissions along with imported and exported products. The purpose of these taxes would be to internalize the external cost of dirty energy such as coal and force coal and power enterprises to use clean coal technologies.

Sixth, China should promote the development of science and technology. Technical innovation is crucial to realizing the energy conservation and emissions reduction goals set by the government. China should encourage the import and use of major technologies as the main means of effectively reducing the cost of emissions reduction. China should also encourage international cooperation in creating new green technologies as the key technical means for medium- and long-term emissions reduction. China must protect intellectual property rights and develop climate adaptation technologies for use in such areas as agriculture, industry, construction, water conservation, and environmental protection. Finally, China must develop standards for climate adaptation technologies.[37]

Seventh, the Chinese government should establish regional policies for emissions reduction that reflect the HDIs of the various regions. The eight provinces and municipalities along the coasts, home to 30 percent of the population, belong to the high-HDI group. The new objective for these areas should be to cut 2010 levels of CO_2 emissions by 10 percent by 2015. This cut must be mandatory. According to the division of principal functional zones, the main targets for optimal and key development zones are to

—Optimize the industrial structure.

—Reduce the economy's percentage of heavy industry.

—Raise the service industry's percentage (in Beijing, for instance, the service industry has reached a healthy 73 percent).

—Optimize the energy consumption structure.

—Use primarily external and internal quality energy, clean energy, and new energy.

—Greatly reduce the percentage of coal consumption.

—Reduce CO_2 emissions.

The other twenty-three provinces and autonomous regions, home to 70 percent of the Chinese population, belong to the upper-medium HDI categories. They will be required to implement conditional emissions reduction in the Twelfth Five-Year Plan. These conditions can be set based on the areas' carbon emissions as compared with the national average as well as on their relative HDI gaps as compared with the high-HDI group.

Eighth, as discussed above, China should expand green ecological space and develop the forestry industry.

Ninth, China should increase green investment and actively expand the green "new deal." China is the largest investor in energy. The International Energy Agency predicted in 2007 that China could invest $3.7 trillion in energy from 2006 to 2030, of which three-quarters would go toward the power industry.[38] Such an investment would constitute 23 percent of the world total ($16 trillion). After November 2008 the State Energy Administration approved and initiated the construction of three nuclear power projects, ten 1 million kilowatt generating units (at a cost of 120 billion yuan), and 5,300 kilometers of the eastern section of the second west-east natural gas transmission line (requiring an investment of 300 billion yuan).[39] According to estimates by the U.S. Energy Department, China's clean energy market could be worth $186 billion by 2010 and $555 billion by 2020.[40]

Tenth, China should support and develop green trade, restrict the export of energy-intensive and carbon-emissions-intensive products, and eliminate disguised subsidies.[41]

Eleventh and finally, the Chinese government should enhance international cooperation. China should take an active part in assessing and combating global climate change. China should also take a leadership role among developing countries in reducing emissions and breaking the deadlock in climate change negotiations.[42] China should accept and implement the international framework covenant regarding climate change, actively participate in international cooperatives that promote the development of new energy, and provide funds for establishing an international R&D grant for technology transfer.

As Deng Xiaoping says, "our next generation will be smarter than us, and they will find a smart way to solve the problem."[43] China's new leaders will be more creative than the old ones. These new leaders should make a public commitment to reduce CO_2 emissions. As they say, a good beginning is half the success. China will develop along a scientific and green path, contributing to global sustainable development by becoming a model for the achievement of a successful green revolution.

CHAPTER EIGHT

Assessing China's Development Goals and Grand Strategy

This decade (2011–20) will be a crucial period for realizing the goals of the third stage—the first fifty years of the twenty-first century—of China's socialist modernization strategy. This represents the final stage of Deng Xiaoping's three-step strategy proposed in 1978:

—Between 1981 and 1990 China will double GNP and basically solve the problem of food and clothing.

—By the end of the twentieth century GNP will reach $1 trillion, with a per capita income of $800 to $1,000.

—By 2050 China will approach the level of developed countries and basically realize its goal of modernization.

This third stage represents a creative phase in the development of socialism with Chinese characteristics as well as a strategic opportunity for an invigorated Chinese nation to rise and become a strong and prosperous country. During this time, China not only will have great opportunities but will also face challenges it has never before encountered. To build a well-off society along all dimensions and construct a harmonious socialist society, China must continue the path of scientific development. It is therefore necessary to carefully design strategic development goals for the second step on the basis of the Eleventh Five-Year Plan and in line with world development trends and the PRC's unique national conditions.

Basic Principles and Methods for Goal Design

China's economic and social development goals for 2020 should be structured around the following four principles: people-centered scientific development, modern socialist development, millennium development goals, and long-term development goals.

PEOPLE-CENTERED SCIENTIFIC DEVELOPMENT

First, China should establish a people-centered scientific development approach as the basic principle and guide. In promoting the sustainable, rapid, harmonious, and healthy development of the national economy, building a well-off society, and constructing an overall harmonious socialist society, it is necessary to stick to the people-centered scientific development approach that is in compliance with China's development stage and actual national conditions. *People centered* refers to the fact that state functions should take the rights of all of its citizens as its primary concern. On-the-ground work must be done to ensure the basic rights of the people. These include the right to democratic elections, the right to freedom, the right to employment, the right to education, the right to social security, and the right to all-around development. All of these rights are enumerated in the constitution of the People's Republic of China.

Development should be focused on people's core, fundamental, long-term interests and aim to solve the problems that have a direct and practical impact. The people should be made the real masters of the country and the main players in its development. The motto of building a well-off society is development by the people and for the people, with fruits of development shared by the people. The process of constructing a harmonious society requires national effort, with each individual taking part, performing his or her duty, and making his or her contribution.

MODERN SOCIALIST DEVELOPMENT

Second, China should underscore the core idea of building a modern socialist society with Chinese characteristics. China's 2020 goals for economic and social development constitute a grand blueprint for the country's long-term development. For the people of China, it illuminates the socialist society that the PRC will construct over the course of the next ten years. For the world, it explicates what kind of country China will be as it rapidly modernizes and internationalizes, what positive role it will play in the international community, and what influence it will exert on neighboring countries, the Asian region, and the world as a whole. I define a modern socialist society with Chinese characteristics as one in which the following will happen:

—The elements of modernity will steadily increase; major progress will be made in modernization, with education, science and technology, and information modernized first; China will soon reach midlevel in terms of modernization.

—The factors of socialism will steadily intensify; major progress will be made in building a socialist society commensurate with the midlevel of modernization.

—Distinctly Chinese elements will be tapped, innovations which will exert unparalleled influence on world peace, development, and cooperation.

When viewed from a historical point of view, a modern socialist society with Chinese characteristics is far removed both from the traditional agrarian society that was China for several thousand years and from the society of the bourgeois democratic revolution period. It is also different from the capitalist society of Western countries, from the society that dominated developing countries over the course of the past century, and from the socialist model of the former Soviet Union and Eastern European countries, less-developed countries, and emerging industrialized countries. Compared to these, China will be more open and more tolerant, but taking on the best aspects of each, such as the modern market economy, democracy, the rule of law, human rights, and freedom. It will sinocize (*zhonguohua*), or localize, them; that is, modernization will take on Chinese characteristics, culture, and innovations. The "well-off" level of life put forward by Deng Xiaoping, the "well-off society" visualized by Jiang Zemin, and the concept of "harmonious society" and "peaceful development" developed by Hu Jintao are all modern Chinese innovations that bear strong Chinese and socialist characteristics.

Millennium Development Goals

Third, China should integrate three major goals and make them the country's future development goals.[1] The three major goals are:

—Building a well-off society in a comprehensive manner, as explicated by the Sixteenth Party Congress.

—Constructing a harmonious socialist society, as announced at the sixth plenary session of the Sixteenth Central Committee of the CCP.

—Pursuing the Millennium Development Goals (MDGs) put forward by the international community. These three goals are multidimensional, interrelated, and complementary. Their multidimensionality refers to the fact that they involve not one indicator but multiple indicators and not development indicators in a single area but development indicators in many areas. For instance, the goal of building a well-off society involves GDP, the economy's structure, economic efficiency, overall national power, international competitiveness, industrialization, the urban population, the development gap, social security, employment, people's livelihoods, democracy and the legal system, social order, population qualities, education, science and technology, culture, medical and health services, development sustainability, ecology and environment, and the utilization rate of resources.[2] The goal of constructing a socialist harmonious society put forward at the sixth plenary session of the Sixteenth Central Committee includes nine major indicators.[3] The MDGs include eight major goals, eighteen subgoals, and forty-eight indicators.[4] These goals and indicators are interrelated. The realization of one goal will have an impact on the pursuit of the other goals. Their being complementary means that the goal of building a well-off society and the goal of constructing a harmonious society are mutually reinforcing. Domestic goals and international goals are also integrated and complementary. The goals of building a well-off society and a socialist harmonious society are the Chinese version of the MDGs. Moreover, in building a well-off society and a harmonious society China will contribute to the implementation of the MDGs.

On this basis, this book has generalized seven major goals for economic and social development up to the year 2020 and then, according to the sequence of priority, identified twenty-two indicators of top priority and twenty-eight indicators of secondary priority, totaling fifty quantified and time-bound development indicators. (Details are discussed below.)

Long-Term Development Goals

Fourth, China should ensure that the goal design fully embodies the detailed views, strategies, and policies of long-term development. In future development, efforts will mainly be concentrated on implementing the goals in rural areas, where low-income segments of the population and vulnerable groups are concentrated. Steps will also be taken to narrow the gap between the city and the country, among regions, and among groups of people in such areas as income, employment, education, health, the provision of basic rights, insurance, and basic public services. There will also be a focus on raising these areas' human capital level and development capabilities. Together, the combination of these efforts will unify the goals of building a well-off society and constructing a harmonious socialist society, bringing together the economic, social, human development, and domestic development goals with the MDGs. The goals are realistic and reachable, with room to spare, and they are both qualitative and quantitative. They follow on the tail of short-term goals of the Eleventh Five-Year Development Program, taking in all the innovations of the period before them. And they provide an important basis for formulating the intermediate goals (that is, the Twelfth Five-Year Development Program).

Socialist Modernization by 2020

This book outlines the major goals for economic and social development by 2020 based on China's overall national conditions, development stage, trends, conditions, and challenges. These goals are also in line with the general requirements for building a well-off society and constructing a harmonious socialist society.

ECONOMIC GROWTH AND STRUCTURAL GOALS

Economic growth and structural goals include the following seven aspects:

—Realize industrialization.[5] In the next decade, the GDP will average an annual growth rate of 7.5–8.0 percent. This means that total GDP in 2020 will be 5.5–5.8 times that of 2000, fulfilling or exceeding the goals set by the Sixteenth Party Congress.

—Maintain stability of the macroeconomy. The Chinese economy should prevent both large rises and large falls and small rises and small falls. The general price level should be kept basically stable, with fiscal receipts and expenditures and international payments roughly balanced.

—Follow a new road of industrialization. Information should be utilized to stimulate industrialization, and industrialization should in turn be used to stimulate the development of the information industry.[6] It is therefore imperative to optimize the industrial structure, reorganize and transform traditional industries, develop specialty industries, stimulate high-tech and high-value-added industries, and raise the proportion of the service industry in the production and employment structures.

—Realize almost full employment. China's abundant labor resources should be fully utilized in order to stimulate an appropriate level of employment growth in the working population and to bring about rapid change in the employment structure such that the proportion of nonagricultural workers reaches about 70 percent and the real unemployment rate falls below the natural unemployment rate (5 percent).[7]

—Make major progress in urbanization. Efforts should be made to open up new spaces of development for the rural population, bringing the urban population to over 60 percent, with the Yangtze River delta and the Pearl River delta, as well as the Beijing, Tianjin, and Hebei areas, able to boast the largest population density in China and the world.

—Establish a vital socialist market system. Establishing a nationally unified open and competitive market system will stimulate the free movement of all production factors, especially population and labor, so as to create a pattern of

Table 8-1. *Human Development Indexes, China and the World, 2003*
Units as indicated

Index and group	Mean life expectancy at birth (years)	Adult literacy rate among 15-year-olds and above (percent)	Enrollment, all school levels (percent)	Per capita GDP/ PPP ($)	Human development index	Total population (million)	Share of world total (percent)
High HDI	78.0	n.a.	91	25,665	0.895	1,211.5	19.2
Middle HDI	67.2	79.4	66	4,474	0.718	4,205.8	66.6
Low HDI	46.0	57.5	46	1,046	0.486	788.7	12.5
World	67.1	n.a.	67	8,229	0.741	6,313.8	100.0
China	71.6	90.9	69	5,003	0.755	1,300.0	20.6

Source: UN Development Program, *Human Development Report 2005.*
n.a. = Not available.

regional cooperation and common development. Such a pattern should feature specialized divisions of labor with distinct characteristics. These divisions, however, should also be mutually complementary and beneficial.

—Build an economic system that is open and competitive, ensuring the opening of neighboring, regional, and multilateral free trade zones, thus expanding the geographical space for China's development and sustaining China's status as the number-one exporter in the world. China should also edge into the front ranks in terms of trade in services and become a power in foreign investment.

COMMON PROSPERITY GOALS

"To make people wealthy" means to strive to make all of the Chinese people wealthy, which is the most important part of building a well-off society. This objective can be broken down into five aspects:

—Raise China's HDI to the upper middle level and high level (above 0.80) and advance the country's world ranking. Its HDI should be raised from 0.76 (already above the world average of 0.74) in 2003 to 0.88 by 2020 (the average of the high-HDI group in the world is 0.90 (table 8-1). This will raise China's world ranking from eighty-fifth among 177 countries in 2003 to about fiftieth.[8] The number of people in China with a high level of HDI (above 0.80) will increase from 280 million (22 percent of the total population) in 2003 to over a billion (three-quarters of the total population) by 2020, amounting to five-sixths of the world's 1.2 billion people in the high-HDI group.[9] This is an important hallmark of a well-off society and would also represent a major contribution to human development in both China and the world as a whole.

—Further raise per capita income levels and significantly advance China's world ranking in this respect, edging into the middle-income group. According to the exchange method, China's per capita GNI will be brought to $8,000–$8,500 from $4,000 in 2010, with its world ranking rising from 110th of 180 countries to about 80th by 2020.[10]

—Eliminate extreme poverty in rural areas. The number of people with daily expenditures of less than one dollar will be reduced to less than 5 million by 2020, from 149 million in 2005, and the number of people living in relative poverty (spending less than two dollars daily) will be halved. The system of minimum cost of living will cover both urban and rural areas. In addition, charities, donations, mutual aid, and volunteers will be encouraged to help these groups.

—Improve the health-for-all indicators. The mean life expectancy at birth will be raised from seventy-two years in 2005 to seventy-six years in 2020. The child mortality rate will be reduced by one-third, from 2.33 percent in 2005 to 1.55 percent. The mortality rate of women during childbirth will be reduced by one-third (from 0.45 percent in 2005 to 0.30 percent). AIDS and communicable diseases will be brought under control. The rural population will be covered by the primary health service. The use of tap water in rural areas will be raised to 90 percent by 2020, up from 61.7 percent in 2005. The use of sanitary toilets in rural areas will be raised to over 90 percent by 2020, up from 53.1 percent in 2005. Safety problems associated with the drinking water used by more than 300 million rural people (90 percent in the middle and western areas) will be resolved.[11] Government investment in public health and basic medical services will be raised to about 7 percent of GDP.[12]

—Address the widening gap between urban and rural areas and among regions step by step.

EDUCATION AND SCIENCE AND TECHNOLOGY GOALS

China should aim to achieve the following by 2020: modernizing education and continuing to improve in science and technology.

Modernize Education

Fiscal expenditures on education will be raised from 2.8 percent of GDP in 2005 to 4.0 percent by 2012 and to 6 percent by 2020. Fiscal expenditures on elementary education will be raised from 0.92 percent of GDP in

Table 8-2. *School Enrollment, China, Selected Years, 1990–2020*
Percent

Student body	1990	2000	2005	2010	2020
Junior secondary school, 12–14-year-olds	66.7	88.6	95.0	99.0	99.5
Senior secondary school, 15–17-year-olds	24.0	41.0	52.7	80.0	90.0
Tertiary school, 18–22-year-olds	3.4	12.5	21.0	26.0	40.0

Source: *China Statistical Abstract 2007,* p. 179; Ministry of Education of the People's Republic of China, *Eleventh Five-Year Program for Education* (2007); author's calculations.

2005 to 1.4 percent by 2012 and to 2.0 percent by 2020.[13] Elementary and junior secondary education will be made compulsory by 2010, eliminating illiteracy among the young and the middle-aged.[14] All locales will be encouraged to extend compulsory education to age twelve and to bring the gross enrollment of senior secondary schools from 53 percent in 2005 to 90 percent by 2020 (table 8-2).[15]

Intermediate vocational education will be expanded on the same scale.[16] Tertiary school enrollment will be raised from 21 percent in 2005 to over 40 percent by 2020, and the number of people receiving postgraduate education will increase from 61 million in 2005 to about 180 million by 2020, with the portion the total population receiving such an education rising from 5.2 percent in 2005 to over 12 percent by 2020.[17] The average years of education attained by people over age fifteen will be raised from 8.5 years in 2005 to over 10.0 years in 2020. The years of education for increased labor will be raised from 10.0 years in 2005 to over 13.5 years in 2020. The percentage of the workforce with higher education will rise from 8.5 percent in 2005 to 20.0 percent in 2020.

China has already surpassed the United States in the number of students in college, and it will soon surpass the United States in the number of students working for master's degrees and doctorates. The biggest disadvantage China faces vis-à-vis the United States is in quality of education. In the future, China should try to narrow its gap with the United States in the stature of its universities and in training high-level professionals. It should build up a number of universities with internationally acknowledged high standards, excel in several of the world's key academic fields, and train millions of outstanding personnel to compete internationally. All of this can be accomplished by reforming the education system and increasing investment in it.

Modernize Science and Technology

China should build the largest scientific and technological community of scientists and technology experts in the world.[18] The number of personnel engaged in these fields should grow from 32 million to 60 million by 2020, with this group as a percentage of total employment rising from 4 percent to 8 percent. Increased investments should be made in R&D, with its proportion of GDP rising to over 2.5 percent by 2020. China's rise to be one of the world's scientific and technical leaders will increase science and technology's level of contribution to global economic growth to over 60 percent. China will rank among the top five in the world in terms of both the number of patents granted to its citizens and the number of research papers quoted globally.[19]

The proportion of China's GDP contributed by high-tech industries will be raised from 4.3 percent to over 10 percent. In addition, the proportion of high-technology exports that originate in China will rise, intellectual property rights will be strengthened, and high-value-added products will become more prominent, making China one of the world's most competitive and largest high-tech production, manufacturing, design, R&D, and export bases.[20] The diffusion, transformation, trade, and application of research achievements should be accelerated, and the domestic technology market trading should be encouraged, with its transaction value as a proportion of GDP rising from 0.85 percent to over 2.5 percent.

By 2020 China will become the world's major S&T power. From 1980 to 2007 changes in the proportions of the five major S&T indicators in the world total suggest that the relative gap between China and developed countries such as the United States and Japan is rapidly narrowing. The S&T power gap between China and the United States (calculated using the five major weighted indicators) was narrowed from 32 times to 2.4 times between these years. The gap between China and Japan was narrowed from 20.4 times to 1.5 times. Looking forward, the United States and Japan will remain China's targets—the countries China wants to catch up to in terms of S&T power—and the relative gap between the PRC and these countries will continue to narrow. By 2020 at least two goals will be realized: China will catch up with and surpass Japan to become the world's second-biggest S&T power; and the gap between China and the United States will be narrowed to two times, with the PRC passing the United States in terms of computer users.

Sustainable Development Goals

There are three goals for China's sustainable development:

—Build a resources-efficient society. By 2020 energy consumption per unit output will be halved. Total water use will remain unchanged, while water used by agriculture will be further reduced.[21] The water utilization rate will be raised by over 20 percent, and the effective coefficient of irrigation water will rise from 0.45 to 0.55.

—Build an environment-friendly society. By 2020 total investment in environmental protection will make up 2.5 percent of GDP, almost double its current proportion (1.3 percent). The discharge of main pollutants, such as sulfur dioxide, nitric oxide, suspended particles, and acid rain, will be reduced by over 33 percent. Forest coverage will reach 23–24 percent. The standing stock of timber will increase by 30 percent. The areas of land dedicated to nature reserves will increase.[22]

—Improve the disaster reduction and relief system. Natural disasters are inevitable, but disaster reduction and relief equates to increases in GDP and the overall welfare of the nation. The safety net against natural disasters should be strengthened and an emergency response and relief mechanism should be set up. In addition, nationwide insurance against disasters should be introduced to mitigate the damage caused by natural disasters. At present, the central government has already started to implement agriculture disaster insurance and catastrophe house insurance in several provinces as pilot projects.[23]

Social Harmony and Stability Goals

The three goals for maintaining social harmony and stability are as follows:

—Set up a basic system of social security by 2020. Over 85 percent of urban residents should be covered by unemployment insurance, medical insurance, and basic old-age insurance. Additionally, the coverage rate of rural medical insurance should reach 100 percent.

—Make society more secure. Labor disputes should be constrained to five per 10,000 of total population; criminal cases should be controlled within about six per 10,000 of total population; challenging the public order cases should be controlled at about eight per 10,000 of total population. The state will intensify public security, improving its oversight as regards disaster reduction and relief, safety production levels, the death rate per unit of GDP, the safe use of food and medicine, state security, and social stability. It will also establish an emergency system for responding to breaking events.

—Create a civil society defined by a high level of civility, social ethics, and harmony nationwide.

Democracy and Rule of Law Goals

Democracy and rule of law targets include the following:

—Improve the system of socialist democracy. The National People's Congress system should be reformed and improved, and the multiparty cooperation and political consultation system led by the CCP should be improved. The broadest possible united front should be consolidated and expanded. The system of regional autonomy should be improved, and grassroots democratic management in urban and rural areas should be perfected. Democratic supervision and more transparent government should be introduced.

—Set up and improve the socialist legal system. A law system with Chinese characteristics should be implemented, and a clean government system should be established and perfected.

Building a Strong, Modern, Socialist Country

There are four specific goals for building a strong, modern, socialist country:

—Raise China's economic aggregate to the front ranks of the world. If calculated using the exchange rate method, China's GDP is the second largest since 2010. If calculated using the PPP method, China's GDP will rise from second largest in 2005 to largest by 2020 (table 8-3).

—Enhance international competitiveness. China will move from its current ranking of between thirtieth and fortieth into a ranking of between tenth and fifteenth in the world. Behind this improvement will be the emergence of a number of enterprises commanding core technologies and boasting internationally known brands and international competitiveness. The number of Chinese enterprises among the world's top 500 will increase from 19 in 2006 to more than 120 in 2020.[24]

—Build up a powerful modern army to safeguard state security and national unity and to prevent and contain Taiwan from being separated from the motherland.

—Enhance integrated national power. The integrated national power of China and the United States will further rise in terms of their percentage of the world total, and the gap between the two powers will be narrowed from 1.5 times in 2010 to about 1.2 times by 2020.[25]

The goal of national development for 2020 is to "build a wealthy and powerful, democratic, civilized and harmonious socialist modernized country."[26] This goal's indicator system and major tasks contain seven major aspects in

Table 8-3. *Major World Economic Indicators, China, Selected Years, 1978–2020*
Units as indicated

Indicator	1978	1990	2000	2005	2010	2020
GDP[a]						
Ranking	10	11	6	4	2	2
Percent of world total	1.7	1.6	3.8	4.9	10.0	n.a.
GDP[b]						
PPP ranking	4	3	2	2	2	1
Percent of world total	4.9	7.8	11.8	15	18	22
Imports and exports						
Ranking	29	16	8	3	2	2
Percent of world total	0.9	1.7	3.6	6.7	10	> 15
Foreign Exchange Reserve	40	7	2	2	1	1
Science and technology ranking	n.a.	n.a.	5	3	3	2
Comprehensive national power[c]						
Ranking	5	3	2	2	2	2
Percent of world total	4.7	5.9	8.7	11.5	14.2	n.a.

Source: *China Statistical Abstract 2007;* World Bank, *World Development Indictors 2006;* Angus Maddison, "Statistics on World Population, GDP, and per Capita GDP, 1–2006 AD" (www.ggdc. net/maddison); World Trade Organization, *International Trade Statistics,* various years; Ministry of Science and Technology of the People's Republic of China, *S&T Statistics Data Book 2010;* Hu Angang and Xiong Yizhi, "Quantitative Assessment of China's S&T Power, 1980–2004," *National Conditions Report,* no. 22 (June 2006); Hu Angang and Wang Yahua, *National Conditions and Development* (Beijing: Tsinghua University Press, 2005), p. 17; author's calculations.

n.a. = Not available.

a. GDP is calculated by the exchange rate method, calculated using the 2000 dollar.

b. GDP is calculated by the PPP method and 1990 international dollar.

c. Comprehensive national power covers eight strategic resources and twenty-three indicators.

the domains of "economic growth, people's wealth (human development), science, technology, and education, sustainable development, harmony and stability, democracy and rule of law, and soft power." Together, these indicators constitute a road map for China's modernization during the first twenty years of the twenty-first century. They also represent the integration of long-term and short-term, future and practical, extensive and specific, and bold and feasible goals. These are the people-centered goals, such as poverty reduction, health improvement, and environmental protection, that will directly improve social development. They will also enhance development in areas such as investment in education, training, skills, health, basic R&D, infrastructure, and social security, which are aimed at increasing productivity and developing new productive forces.

Highlighting Priorities for 2020

According to the major goals of national development, which take the numbers of 2010 as their base, twenty-six development targets have been identified as top priorities (table 8-4). These goals have the following five features:

First, the goals reflect the core interests of the state and the priorities of national economic and social development. They are also an expression of the fundamental interests of all the Chinese people. These goals are observable, quantifiable, comparable (historically, regionally, and internationally), assessable (medium term and ex post facto), and have a clear time limit (five years or fifteen years).

Second, the goals mainly focus on public services, with economic indicators as a supplement. This focus reflects the major changes to and transition of government functions. There are only four indicators for economic growth and economic structure, accounting for 15.4 percent of the total. In the Eleventh Five-Year Development Program, there are six, accounting for 27.3 percent.[27] The recommendation issued at the sixth plenary session of the Sixteenth Central Committee cuts economic indicators and adds indicators for economic activities, such as the four indicators on resources and environment, three on science, technology, and education, three on poverty reduction and public services, and four on population and living standards. It is the first time that four indicators on social harmony were included as top priority.

Third, the targets mainly feature mandatory indicators that each level of government must follow and reach, with supplementary anticipatory indicators. This is done to give prominence to the responsibilities of the government for providing public services. By following the indicator designing method used in the Eleventh Five-Year Development Program, I divided the major development targets into anticipatory and obligatory indicators. Anticipatory indicators are what the state expects to achieve, mainly by the actions of market players. It is up to the government to create good macroeconomic, institutional, and market environments, to regulate the macroeconomy at the appropriate times, and to employ multiple economic policies to guide the allocation of resources. Obligatory indicators are the requirements set by the central

Table 8-4. *Development Objectives, China, Selected Years, 2000–20*[a]
Units as indicated

Indicator	Nature	2000	2005	2010	2020
Economic growth and economic structure					
GDP growth (percent)	Anticipatory	(8.6)	(9.5)	(11.1)	(7.5)
Urban new employment (million)	Obligatory	(41.1)	(41.8)	(45.0)	(100.0)
Transferred rural labor (million)	Obligatory	n.a.	(40.0)	(45.0)	(120.0)
Urbanization (percent)	Anticipatory	36.2	43.0	47.0	> 55.0
Resources and environment					
Reduction of unit GDP energy consumption (percent)	Obligatory	n.a.	n.a.	20.0	20.0
Ratio of renewable energy consumption (percent)	Obligatory	n.a.	5	10	17
Reduction of unit GDP CO_2 emission (percent)	Obligatory	n.a.	n.a.	n.a.	20
Arable land (million hectares)	Obligatory	1.32	1.22	1.20	1.18
Unit GDP water consumption (M3/10,000 yuan)	Obligatory	515	407	255	117
Forest cover (percent)	Obligatory	16.55	18.2	> 20	23.4
Major pollutants (percent)	Obligatory	n.a.	n.a.	(10)	(15–20)
Science, technology, and education					
Patent grants (10,000)	Anticipatory	9.52	17. 16	30	80
Average years of education	Anticipatory	n.a.	8.5	9.0	> 10
Junior secondary school enrollment (percent)	Obligatory	88.6	95.0	97.0	99.0
Rural compulsory education	Obligatory	. . .	. . .	100	100
Poverty reduction and public service					
Rural poverty and low-income people (million)	Obligatory	. . .	14,900	6,000	> 500
Rural people drinking unsafe water (million)	Obligatory	. . .	300	60	> 0. 2
Children planned immunization (percent)	Obligatory	85.3	87.0	> 95.0	100

(continued)

Table 8-4 (*continued*)

Indicator	Nature	2000	2005	2010	2020
Measures of prosperity					
Mean life expectancy (years)	Anticipatory	71.4	72.0	74.5	77.0
Urban disposable income growth (percent)	Anticipatory	(5.7)	(9.7)	(9.3)	(5)
Rural disposable income growth (percent)	Anticipatory	(4.7)	(5.3)	(7.9)	(5)
Human development index (percent)	Anticipatory	0.73	0.76	0.80	0.88
Harmonious society					
Public order cases (1 per 100,000)	Obligatory	469	750	700	600
Urban medical insurance coverage (percent)	Obligatory	31	50	> 60	100
Urban old-age insurance coverage (percent)	Obligatory	45	48	> 55	> 95
Rural medical insurance coverage (percent)	Obligatory	3.1	23.5	100	100

Source: *China Statistical Abstract 2006; Eleventh Five-Year Development Program;* author's calculations.
n.a. = Not available.
a. Annual growth rate, or accumulated growth rate, or accumulative amount is shown in parentheses.

government for local governments and central government departments in areas such as public services. They are to be realized by the government by rationally allocating public resources and effectively employing administrative forces. Among the twenty-six top priority indicators, eight are anticipatory, accounting for 30.8 percent. During the Eleventh Five-Year Development period, fourteen indicators were anticipatory, accounting for 63.6 percent. I recommend adding sixteen obligatory indicators, accounting for 61.5 percent of all targets. This would strongly intensify the responsibilities of the government in the provision of basic public services.

Fourth, the targets for providing public services to rural areas are given special prominence. This indicates that the emphasis in building a well-off society is on the rural areas and that the primary beneficiaries of basic public services are peasants. There are six targets directly concerning the provision of public services to peasants, accounting for 27.3 percent of the total, along with five (or 22.7 percent) indirectly aiding peasants. The combined total of the two makes up 50 percent of all objectives.

Fifth, twenty-seven indicators are of secondary importance and are supplements to the twenty-six top-priority indicators (table 8-5). These indicators cover the major aspects of economic and social development, especially public services. Only four are indicators for economic growth and the economic structure, all of which are expectative indicators and account for 14.8 percent of the total, while twenty-three are noneconomic indicators, which include sixteen obligatory indicators and account for 59.3 percent. The rankings of the indicators are not set in stone. Rather, they are adjustable and interchangeable. When the priority indicators such as GDP growth are easily achieved, then the secondary indicators are implemented more strictly and in a way reasonable to pursue and achieve; when the indicators of secondary importance directly affect the immediate interests of the majority of the people, such indicators can be implemented as priority indicators.

The Seventeenth Party Congress designed and made arrangements for the second-step strategy for the first half of the twenty-first century and prepared to proceed with the third step. I believe that China is very likely to realize the second-step strategic vision of building a well-off society, constructing a socialist harmonious society, and eliminating absolute poverty.

China enjoys many favorable conditions for realizing the above objectives and has become increasingly capable of implementing new policies to attain its goals. Sustained high growth has boosted China's economic aggregate to 18.6 times that of 1978.[28] It will be doubled and redoubled in the future. China's growth in comprehensive national power has made it possible to narrow the gap between China and the United States.[29] The government has also improved its ability to draw on financial resources; the proportion of fiscal revenue in GDP has increased from 10.7 percent in 1995 to 20.4 percent in 2008.[30] The financial resources available for use in reducing poverty, providing social security, and providing various public services will continue to increase. People's lives have improved markedly. Urban and rural private spending on housing, communications and telecommunications, education, culture activities, and health services has increased to varying degrees.

The Chinese people desire stability, development, and harmony. The pattern of opening up all areas of endeavor is still taking shape. China is becoming more and more capable of drawing on foreign resources and technology. The international community has universally been impacted by the huge market and broad development prospects of China. All of these developments will provide the materials as well as the economic and financial conditions for the realization of the above goals.

Table 8-5. *Secondary Development Objectives, China, Selected Years, 2000–20*

Indicator	Nature	2000	2005	2010	2020
Economic growth and macroeconomic control					
Per capita GDP (dollars per person)	Anticipatory	949	1,731	4,045	8,500
Urban registered unemployment (percent)	Anticipatory	3.1	4.0	< 5.0	< 4.0
Employment of service trade (percent)	Anticipatory	27.5	31.4	36.0	> 40.0
Total foreign trade (billions of dollars)	Anticipatory	47.43	142.19	2,800	> 5,000
Resource and environment					
Standing stock of timber growth (percent)	Obligatory	n.a.	> (10)	> (20)	> (30)
Spending on pollution control (percent of GDP)	Obligatory	1.0	1.3	1.8	> 2.5
Science, technology, and education					
R&D spending (percent of GDP)	Anticipatory	1.0	1.3	1.8	> 2.5
Enterprises in Fortune 500 (number)	Anticipatory	9	15	> 50	> 120
Fiscal spending on education (percent of GDP)	Obligatory	2.58	2.79	3.84	> 5
Enrollment senior secondary schools (percent)	Anticipatory	42.8	52.7	70	95
Training (number of migrant workers)	Obligatory	428,000	527,000	700,000	850,000
Secondary vocational education (millions of students)	Obligatory	5.03	6.26	8	10
Public service					
AIDS patients (number)	Obligatory	600,000	650,000	1.5 million	< 3 million
Rural infant mortality rate (percent)	Anticipatory	35	30	12	< 8
Rural maternal mortality rate (1/100,000)	Anticipatory	70	50	30	< 15
Villages accessible by highway (percent)	Obligatory	n.a.	96	100	100
Rural sanitary latrines (percent)	Obligatory	44.8	53	65	> 90
Rural tap water use (percent)	Obligatory	55.2	61.7	70	> 90

Table 8-5 (*continued*)

Indicator	Nature	2000	2005	2010	2020
Rural telecom and TV penetration (percent)	Obligatory	n.a.	> 90	100	100
Death rate from work injuries per unit GDP (percent)	Obligatory	n.a.	n.a.	35	> 70
Population and children					
Total population (millions)	Anticipatory	1,267.43	1,307.56	1,344.97	1,420.80
Gender ratio of children ages zero to four	Anticipatory	n.a.	122.7	. . .	. . .
Rural Engel coefficient	Anticipatory	49.1	45.5	40	> 33
Harmonious society					
Gini coefficient	Anticipatory	0.45	0.46	0.45	< 0.4
Criminal cases (1 per 100,000)	Obligatory	340	500	550	450
Labor disputes (1 per 10,000)	Anticipatory	2	3	3.2	3.0
Urban unemployment insurance coverage[a]	Obligatory	45	39	> 55	> 95

Source: *China Statistical Abstract 2005; Eleventh Five-Year Development Program;* author's calculations.
n.a. = Not available.
a. Calculated by number of people employed.

Conclusion

China's rise, or whether or not China is able to emerge as a superpower by 2020, is a scenario still rife with uncertainty, mainly stemming from international and domestic challenges. Chinese leaders must work to accurately understand the relations between China and the world as well as try to solve China's various domestic problems, as these are the factors that will undergird China's modern renaissance.

External Challenges to China's 2020 Goals

Currently, the world economy has entered the post-financial-crisis era. Uncertainty, instability, imbalances, and insecurity still exist, and some of these infirmities may worsen. As the international political and economic order undergoes this profound adjustment, China and the world are drawn closer together and their mutual influence increases. Relations between China and leading powers, for example the United States, have changed significantly and have led to the enhancement of China's standing as a world economic power.

At present, China has become the superpower of natural resource consumption and greenhouse gas emissions. China is the world's largest consumer of coal, its largest SO_2 emitter, and one of largest CO_2 emitters. China has also replaced the United States as the world's largest energy consumer, which has served to exacerbate concerns over the supposed China threat. Around the world, excepting large countries such as India, China has become a global target of public criticism on pollution. Moreover, given that the U.S. government has agreed to make carbon emission reduction commitments even if the Chinese government fails to make more positive and proactive commitments to reduce carbon dioxide emissions by 2050, China could soon face even greater international pressure.

In the post-financial-crisis era, developed countries have a reduced appetite for foreign imports, and China therefore is at risk of overcapacity. Although the global economic recovery appears to be on track, it will still require a long period of economic adjustment and rebounding. Governments and citizens in developed countries need to curtail their consumption habits and increase domestic saving rates in order to slow consumption and investment in the medium term and improve their balance sheets. There is mounting evidence that many governments are explicitly or implicitly implementing trade protectionism, which means that in the short term we may witness a degree of deglobalization. Because the conventional imbalance of international trade has been thrown off, China may find it difficult to continue its highly export-reliant economic growth model.

Economic and trade conflicts have increased, both in number and variety, in the countries that are developing economic and trade relations with China. Moreover, since the 1990s the Chinese economy has gradually transformed from one with shortages to one with surpluses, which has caused the domestic economy to become more and more imbalanced. Nevertheless, the large trade surplus has largely offset and covered for internal problems relating to production and overcapacity.

Uncertainty is now increasing in the global system, and these global risks may transform into domestic risks. With adjustments to the global economic development pattern and geopolitical order, the world will remain in a period of significant uncertainty and instability. Traditional and nontraditional threats are closely intertwined. Commodity markets and financial market prices may remain volatile. Trade protectionism and trade disputes have increased. Mutual reliance between the major powers is coupled with mutual competition.

In this context, dependence on external sources of energy and natural resources is also much greater. Since net oil exports turned into net oil imports in China in 1993, China's external dependence ratio for oil has increased steadily. In 2007 this ratio reached 50 percent, and it will continue to increase. From a natural resource perspective, China's domestic resources are relatively scarce; therefore, an increase in China's use of world resources is inevitable. China became a net importer of primary products in 1995 through rising imports. Net imports of primary products in 2006 accounted for 5.11 percent of GDP and may reach 12 percent by 2020.

Closer ties between China and the world have created new challenges to public health. China is already a highly mobilized country. Foreign inbound tourism in 2008 reached 130 million, and domestic outbound visits reached 46 million. China needs to improve its public health system in a transparent way and actively participate in international public health cooperation.[31]

Domestic Challenges to China's 2020 Goals

Since the period of reform and opening, the Chinese economy has maintained a strong and rapid growth for three decades.[32] From 1978 to 2009 the GDP growth rate was 9.9 percent; per capita GDP growth rate was 8.7 percent. To an even greater extent in the new century, China has entered a golden period of accelerated development. For five consecutive years, 2003–07, China's GDP was more than 10 percent. At the same time, this has engendered a period of contradictions in the domains of natural resources, the environment, society, and the global system.

Based on this pace of development, China will transition from a lower middle-income country to a middle-income country. Accounting for purchasing power parity, China's per capita GDP now exceeds the world average and has reached Western European levels of the 1960s and 1970s. At this stage China must avoid falling into the middle-income "trap" and continue to maintain macroeconomic stability, social stability, and political stability, thereby achieving a smooth upward trajectory from middle-income to high-income levels.

The long-term deterioration of the income distribution is the main reason for the middle-income trap. China's current pattern of income distribution highlights the decline of the proportion of household income to gross national income. The proportion first increased and then decreased. The share of household income in gross national income during the period of 1978–82 rose rapidly, to a peak of 62.8 percent in 1982, which suggests that the residents were the direct beneficiaries of economic growth. In 2007, however, the share declined, reaching a figure lower than the 1978 percentage, 44.5 percent, which reflected income growth significantly lower than economic growth, meaning that high economic growth was not necessarily translating into mean income growth. At the same time, disparities in household income increased and the proportion of labor remuneration in the primary distribution declined. Income disparities among sectors and groups also increased. China's income disparities are not only between urban and rural areas but also within urban and rural areas. The rich get richer and the poor get poorer. Polarization of the income distribution has become a serious challenge facing China. Whether it can reverse this trend will bear on not only social fairness and justice but also social stability.

Additionally, China is entering a period of low birthrates and an increasingly aging society ahead of time, while remaining at a relatively low income level.[33] Even so, unsustainable population growth is a major challenge for China at the moment. It is beginning to suffer "Japanese disease" or "East Asian emerging economies disease," wherein birthrates decline just as the society ages rapidly. China is suffering this about twenty years later than Japan, and its decline in the working-age population of fifteen-to-forty-year-olds will occur about thirty years later than Japan's. But the biggest difference between China and Japan in terms of low birthrate and aging is that Japan was rich before the demographic changes, while China is undergoing this transition at the same time that it struggles to get rich. This will change the future of China's economy, moving it from higher growth to lower growth and, therefore, challenging China's long-term development.

At the beginning of reform and opening rapid economic growth entailed serious environmental pollution. During the Eighth Five-Year Plan pollution reached its peak, and during the Ninth Five-Year Plan we began to see green development patterns (high growth and low carbon emissions) as a result of industrial structure adjustments and greater concern for environmental protection. During the Tenth Five-Year Plan, however, China had a serious pollution problem as a result of the economy's reliance on heavy industrial production. With the implementation of the Eleventh Five-Year Plan, pollution has been effectively controlled and emissions of major pollutants have been significantly reduced. In the near future, the key for sustainable development will be whether or not China can continue to adhere to this trend and achieve increasingly green development. China is engaged in the largest process of urbanization and industrialization in human history, which risks injuring a fragile ecological environment through grassland degradation, desertification, destruction of vegetation, soil erosion, salt alkalization, land contamination, and the greatest biodiversity loss the world has ever seen. The gap between humans and nature will continue to expand, as China is still running an ecological deficit, but the overall destructive trend is somewhat contained.

Although international and domestic uncertainty and instability have increased, there have been no fundamental changes to the favorable internal and external conditions for China's economic development. Economic, technological, political, security, and other international and domestic environments are at their best since the founding of the PRC, and China is now facing a historic opportunity for peaceful development. Peace, development, and cooperation are the mainstream trends of the world, and economic globalization and trade and investment liberalization are irreversible. Science and technology are progressing rapidly and becoming more prevalent, and the innovation competition among countries is more prominent. In the years ahead, regional cooperation and global cooperation will be even more common and the means of cooperation even more diverse. Collective action on global challenges will become more important for every country.

At the same time, issues such as climate change, ecological environment protection, regional security, and terrorism are becoming increasingly hard to ignore and increasingly important to address. This confluence of events will require that we understand how to coordinate domestic and international situations. It is not only necessary to understand the development trends of today's world but also to ascertain China's role in the world. China needs to have overall plans for its long-term development and should share its development opportunities with the world, so that China and the world can respond to these challenges together.

China's rapid rise has changed not only the pattern of world power but also China's role in the international system. This has opened up an unprecedented historic opportunity for China to actively participate and provide leadership in economic globalization, economic integration, trade liberalization, and investment liberalization and has allowed China to further expand its international development space and its access to important international resources. China can promote the reform of global governance systems, break the monopoly of the United States, and assert a greater influence in the world. This can also serve to break the Western culture's long-standing monopoly over modernity and bring more diversified cultures and values to the world stage. China can assume a leadership position in the launch of the fourth industrial revolution (that is, the green industrial revolution) by advocating for an innovative green development model in international negotiations on climate change. This would be the first instance of China actively participating in developing a new pattern of world behavior.

RETHINKING THE CHINA DREAM

What is the China dream? Mao Zedong said fifty years ago that in the twenty-first century China should make a greater contribution to mankind, and in contribution to Hu Jintao's even greater contribution, it is clear that China's position has changed both in comprehensive national power and in other adaptations and rational changes. The question remains: What will be China's contribution to the world?

Besides its enormous economic and trade contributions, as well as poverty reduction contributions, China needs to contribute in four other key areas: human development, science and technology, the green movement, and culture. These four contributions would represent China's modern renaissance, with domestic and international significance. Given that China has become one of the largest stakeholders in world affairs today, it is incumbent upon it not only to follow its own interests and the interests of developing countries but also to develop in a fashion consistent with the interests of developed countries.

Notes

Introduction

1. Robert Gilpin, "The Theory of Hegemonic War," *Journal of Interdisciplinary History* 18, no. 4 (1988): 591–613; and Robert Gilpin, *The Political Economy of International Relations* (Princeton University Press, 1987).

2. The recent English works that offer different assessments of China's rise include Susan L. Shirk, *China: Fragile Superpower* (Oxford University Press, 2008); James Kynge, *China Shakes the World: A Titan's Rise and Troubled Future—and the Challenge for America* (Mariner Books, 2007); Martin Jacques, *When China Rules the World: The End of the Western World and the Birth of a New Global Order* (Penguin, 2009); and Edward S. Steinfeld, *Playing Our Game: Why China's Rise Doesn't Threaten the West* (Oxford University Press, 2010).

3. A few noticeable exceptions include Mark Leonard, *What Does China Think?* (Public Affairs, 2008); Yu Keping, *Democracy Is a Good Thing: Essays on Politics, Society, and Culture in Contemporary China* (Brookings, 2008); and Wang Hui, *China's New Order: Society, Politics, and Economy in Transition* (Harvard University Press, 2006). For collections of writings by Western and Chinese scholars, see Robert S. Ross and Zhu Feng, eds., *China's Ascent: Power, Security, and the Future of International Politics* (Cornell University Press, 2008); and Richard Rosecrance and Gu Guoliang, eds., *Power and Restraint: A Shared Vision for U.S.-China Relationship* (Public Affairs, 2009).

4. Jacques Barzun, "The Man in the American Mask," *Foreign Affairs* 43, no. 3 (1965): 426.

5. Hu Angang, *Zhongguo: Zouxiang Ershiyi Shiji* [China: Toward the 21st century] (Beijing: China Environmental Science Press, 1991), p. 127.

6. Hu Angang, *Guoqing yanjiu yu jiaoshu yuren* [China studies: Teach knowledge and educate people] (Tsinghua University Press, 2011), p. 15.

7. See http://business.sohu.com/85/79/article13737985.shtml.

8. Hu, *Guoqing yanjiu yu jiaoshu yuren*, pp. 89, 165.

9. Ibid., p. 89.

10. For more discussion of the growing role of think tanks and other interest groups in China's policymaking process, see Cheng Li, "China's New Think Tanks: Where Officials, Entrepreneurs, and Scholars Interact," *China Leadership Monitor*, no. 29 (Summer 2009); and Linda Jakobson and Dean Knox, "New Foreign Policy Actors in China," Policy Paper 26 (Stockholm: International Peace Research Institute, 2010).

11. Hu Angang, *2020 Zhongguo: Quanmian Jianshe Xiaokang Shehui* (Tsinghua University Press, 2007).

12. See Hu's speech at the School of Public Policy and Management, Tsinghua University (www.sppm.tsinghua.edu.cn/ggjj/26efe4892607e79901260d56bcbf0007.html).

13. John J. Mearsheimer, *The Tragedy of Great Power Politics* (New York: Norton, 2001).

14. Zhao Yining, *Da Jiaoliang: Dang Zhongguo Long Yudao Meiguo Ying* (Hangzhou: Zhejiang renmin chubanshe, 2010).

15. See http://elite.youth.cn/mj/200903/t20090309_874370.htm.

16. Hu Angang, *Zhongguo fazhan qianjing* [Prospects of China's development] (Hangzhou: Zhejiang renmin chubanshe, 1999), p. 6.

17. *Shijie ribao* [World journal], January 7, 2008, p. A3.

18. Ibid., pp. 5–6.

19. World Bank, *China: Long-Term Development, Issues, and Options* (Johns Hopkins University Press, 1985).

20. See www.un-documents.net/wced-ocf.htm.

21. *China Daily*, July 4, 2010; and September 16, 2010; and also see www.chinadaily.com.cn/zgrbjx/2009-09/16/content_9090404.htm.

22. See http://business.sohu.com/20091208/n268772861.shtml.

23. Hu, *Guoqing yanjiu yu jiaoshu yuren*, p. 101.

24. Ibid., p. 101.

25. For the list of the special reports, see http://ccs.tsinghua.edu.cn.

26. *Renmin ribao* [People's daily], September 5, 2010, p. 11.

27. Hu, "Zhongguo renlei bu'anquan de zuida tiaozhan"; and see also http://business.sohu.com. June 19, 2005.

28. Hu, *Zhongguo fazhan qianjing*, p. 8.

29. Hu, *Guoqing yanjiu yu jiaoshu yuren*, p. 17.

30. For Hu's suggestions, see Zhang Xiaoxia, *Zhongguo gaoceng zhinang* [China's top think tanks], vol. 1 (Beijing: Jinghua chubanshe, 2000), pp. 176–79; and also see www.pinggu.org/bbs/thread-278591-1-1.html.

31. Ibid., pp. 182–83.

32. Joshua Cooper Ramo, *The Beijing Consensus* (London: Foreign Policy Centre, 2004), pp. 22–23.

33. *Liaowang dongfang zhoukan* [Oriental weekly], April 11, 2005.

34. Hu, *Zhongguo fazhan qianjing*.

35. Quoted from Wu An-chia, "Leadership Changes during the Fourth Plenum," *Issues and Studies* 30, no. 10 (1994): 134; Hu, *Zhongguo fazhan qianjing*, p. 312.

36. Hu Angang, *Mao Zedong yu Wenge* [Mao Zedong and the Cultural Revolution] (Hong Kong: Strong Wind Press, 2009); Tang Shaojie, "Ping Hu Angang Mao Zedong yu Wenge" [Comments on Hu Angang's Mao Zedong and the Cultural Revolution], *Ershiyi shiji* [Twenty-first century), no. 116 (November 2009): 113–19.

37. Tsai Wen-Shuen, "Zhonggong zhizheng dianfan de zhuanyi" [The paradigm shifts of CCP's rule] (Taipei, unpublished paper, 2009).

38. Hu Lianhe and Hu Angang, "Zhongguo weihe buneng gao sanquan fenli" [Why China should not adopt the separation of powers], *Renmin ribao* [People's daily], May 10, 2010. For some of the criticism, see www.rfa.org/cantonese/commentaries/weipu-05262010124134.html?encoding=traditional.

39. *Shijie ribao*, February 21, 2010, p. A1.

40. See www.treasury.gov/resource-center/data-chart-center/tic/Documents/mfh.txt.

41. For the online version of *The Economist* article, see www.economist.com/node/17732859.

42. Hu, *Zhongguo fazhan qianjing*, p. 14.

43. See http://finmanac.blogspot.com/2009/02/top-10-banks-in-world-by-market.html.

44. *Caifu* [Fortune], China edition, October 2010, pp. 125–30.

45. *Shijie ribao*, December 30, 2010, p. A3.

46. *Shijie ribao*, February 14, 2010, p. A10.

47. "Trading Places: The World Largest Container Ports," *The Economist*, August 24, 2010. Also see www.economist.com/node/16881727.

48. Ibid. Hong Kong was on the 1989 list of top twenty, but it did not belong to the PRC until 1997.

49. The annual growth rate of output and sales is 48 percent and 46 percent, respectively. Quoted from Zhang Xue, "Domestic Auto Sector Undergoes Structural Adjustments," *Economic Daily*, February 9, 2010.

50. Lu Xueyi, *Dangdai Zhongguo shehui jiegou* [Social structure of contemporary China] (Beijing: Shehui kexuewenxian chubanshe, 2010), pp. 402–06.

51. *Zhongguo qingnian bao*, February 11, 2010.

52. *Zhongguo xinwen zhoukan* [China Newsweek], January 22, 2010. Euromonitor International, a London-based research and consulting firm, predicted that China's middle class would reach 700 million in 2020, about 48 percent of the country's total population. See www.euromonitor.com/Chinas_middle_class_reaches_80_million.

53. Homi Kharas and Geoffrey Gertz, "The New Global Middle Class: A Crossover from West to East," in *China's Emerging Middle Class: Beyond Economic Transformation*, edited by Cheng Li (Brookings, 2010), p. 38.

54. *Shijie ribao*, December 22, 2010, p. A3.

55. *Shijie ribao*, April 9, 2010, p. A1.

56. For more discussion on this topic, see Wang Huiyao, "China's National Talent Plan: Key Measures and Objectives" (www.brookings.edu/papers/2010/1123_china_talent_wang.aspx).

57. See www.chinaelections.org/index.html and www.ftchinese.com.

58. For the main arguments of the China model, see Pan Wei, ed., *Zhongguo moshi: Jiedu renmin gongheguo de liushinian* [China model: A new development model from the sixty years of the People's Republic] (Beijing: Central Compilation and Translation Press, 2009). Hu Angang is a contributor to this edited book.

59. Cai Fang, "A Tale of Two Cities: Chinese Labor Market Performance in 2009 and Reform Priority in 2010," *East Asia Forum,* December 25, 2009; and also see http://news.xinhuanet.com/fortune/2006-08/03/content_4913519.htm.

60. Yu Yongding, "A Different Road Forward," *China Daily*, December 23, 2010. Also see www.chinadaily.com.cn/opinion/2010-12/23/content_11742757.htm.

61. Yu, "A Different Road Forward."

62. See www.iwep.org.cn/Corporation/infoDetail13.asp?cInfoId=177&dInfoId=200.

63. *Beijing shangbao* [Beijing business daily], August 30, 2010. See also http://news.xinhuanet.com/fortune/2010-08/30/c_12496387.htm.

64. See http://xuxiaonian.blog.sohu.com/160724498.html.

65. In his recent book, *The End of the Free Market: Who Wins the War between States and Corporations?* (Portfolio, 2010), Ian Bremmer, the president of the Eurasia Group, also expressed concern about the rise of state capitalism in China.

66. For Xu Xiaoning's views, see http://xuxiaonian.blog.sohu.com/158818651.html. Also *Lianhe zaobao* [United morning news], August 1, 2010 (http://finance.ifeng.com/opinion/zjgc/20100830/2567934.shtml).

67. See http://finance.ifeng.com/news/20101205/3005151.shtml.

68. Sun Liping, "*Zhongguo jinru liyi boyi de shidai*" [China is entering the era of a conflict of interests]" (http://chinesenewsnet.com).

69. *Qianjiang wanbao* [Qianjian evening news], February 11, 2010. Also see www.chinanews.com.cn/estate/estate-lspl/news/2010/02-11/2121577.shtml.

70. See http://bt.xinhuanet.com/2010-03/19/content_19293215.htm.

71. See http://news.xinhuanet.com/fortune/2009-03/17/content_11024848.htm. Zhang Ping, minister of the National Development and Reform Commission, however, told reporters at the National People's Congress's annual meeting in March 2010 that "no penny in the stimulus package has been invested in real estate." See www.sc.xinhuanet.com/content/2010-03/07/content_19178284.htm.

72. Sun Liping, *Duanlie: 20 shiji 90 niandai yilai de Zhongguo shehui* [Cleavage: Chinese society since the 1990s] (Beijing: Shehui kexue wenxian chubanshe, 2003); and Sun Liping, *Zhuanxing yu duanlie* [Transition and cleavage] (Beijing: Tsinghua University Press, 2004).

73. For more discussion on this topic, see Ren Jianming, "Woguo jiaoyu fubai de xianzhang qushi" [Status and trends of China's educational corruption], in *Zhongguo jiaoyu de zhuanxing yu fazhan 2006* [Transition and development of China's education 2006], edited by Yang Dongping (Beijing: Shehui kexue wenxian chubanshe, 2007).

74. Michael G. Finn, "Stay Rates of Foreign Doctoral Recipients from U.S. Universities," paper prepared for the Division of Science Resources Statistics of the National Science Foundation by the Oak Ridge Institute for Science and Education, January 2010, p. 6.

75. Liu Junning, "Jingshen weiji shi zui gengben de weiji" [Crisis in faith is the most fundamental crisis], *Nanfengchuang* [South wind], October 9, 2010.

76. *Jingji guancha bao* [Economic observers], December 25, 2010. Also see www.china-elections.org/NewsInfo.asp?NewsID=195247.

77. Cai Fang, for example, recently gave a presentation at the Politburo meeting.

78. For more discussion on American exceptionalism, see Seymour Martin Lipset, *American Exceptionalism: A Double-Edged Sword* (W. W. Norton, 1997); Deborah L. Madsen, *American Exceptionalism* (University Press of Mississippi, 1998); and Godfrey Hodgson, *The Myth of American Exceptionalism* (Yale University Press, 2010).

79. Henry A. Kissinger, "Avoiding a U.S.-China Cold War," *Washington Post*, January 14, 2011, p. A21.

80. Kang Xiaoguang, "Zhongguo teshulun: Dui Zhongguo dalu 25 nian gaige jingyan de fansi" [Chinese exceptionalism: Reflection on China's 25-year reform], *Confucius 2000*, June 2004. Also see www.confucius2000.com/poetry/zgtsldzgdl25nggjydfs.htm.

81. For an excellent review of the Chinese discussion of Western literature on a rising China, see Liu Yawei, "'Zhongguo teshulun': yiweizhe shenmo?" [What does "China exceptionalism" mean?], *21 shiji guoji pinglun* [21st-century international review] 1, no. 1 (2010): 11–15.

82. Zheng Bijian, *China's Peaceful Rise: Speeches of Zheng Bijian 1997–2005* (Brookings, 2005).

83. Ibid., p. 10.

84. Representative works include Wang Jian and others, *Xin zhanguo shidai* [New era of warring states] (Beijing: Xinhua chubanshe, 2004); and Xiong Guangkai, *Guoji zhanlue yu Xin junshi biange* [International strategy and revolution in military affairs] (Beijing: Tsinghua daxue chubanshe, 2003).

85. Yan Xuetong, "Foreword," in *Zhongguo jueqi jiqi zhanlue* [The rise of China and its strategy], edited by Yan Xuetong and others (Beijing: Peking University Press, 2005), p. 2.

86. Quoted from http://china.dwnews.com/news/2010-12-16/57200778.html.

87. Liu Mingfu, *Zhongguomeng–hou Meiguo shidai de daguo siwei yu zhanlue dingwei* [The China dream: The great power's mindset and strategic stance in the post–American hegemony era] (Beijing: China Friendship Press, 2010).

88. Ibid., p. 21.

89. Ibid., pp. 3–16.

90. See http://news.xinhuanet.com/mil/2010-12/28/c_12926497.htm for Liang Guanglie's remark.

91. Liu Mingfu, *Zhongguomeng–hou Meiguo shidai de daguo siwei yu zhanlue dingwei*, p. 273.

92. Ibid., p. 290.

Chapter One

1. After 1980 the Chinese government mapped out a medium-term (five-year) plan, setting the economic growth target at 7.0 to 7.5 percent and a long-term (ten-to-twenty-year) plan, setting the economic growth target at 7.2 percent. The report to the 1982 Tenth National Party Congress claimed that the industrial and agricultural output value for 2000 would quadruple that of 1980; the report to the 1987 Thirteenth National Party Congress set the target GDP for 2020 as quadrupling that of 2000; the report to the 2007 Seventeenth National Party Congress set the goal of quadrupling the 2000 GDP by 2020. The Tenth Five-Year Plan mapped in 2001 set the target at 7.5 percent, a target that did not change for the Eleventh Five-Year Plan, set in 2006.

2. National Bureau of Statistics, *China Statistical Abstract 2010* (Beijing: China Statistics Press, 2010), pp. 23–25.

3. Computed by the exchange rate method based on the figures released by the World Bank, the U.S. GDP was $14,204.3 billion, 3.3 times that of China; Japan's GDP was $4,909.3 billion, 1.1 times that of China; that of Germany was $3,652.8 billion, 0.84 times that of China. World Bank, *World Development Report 2010: Development and Climate Change*, Chinese ed. (Tsinghua University Press), pp. 374–75. China became the third-largest economy in the world in 2008. In 2010 China overtook Japan as the world's second-biggest economy according to data released on August 16. See "China's Economy: Hello America," *The Economist* (www.economist.com/node/16834943).

4. See World Bank, *World Development Report 1999/2000: Entering the 21st Century* (Oxford University Press). The relative gap in GDP (PPP) between the United States and China narrowed rapidly. That of the United States in 2000 was 1.86 times that of China; the U.S. lead narrowed to 1.17 times by 2006. See Angus Maddison, "Statistics on World Population, GDP, and per Capita GDP, 1–2006 AD" (www.ggdc.net/maddison). By this base figure, China was likely to surpass the United States by 2010; but China's GDP in 2008, according to the CIA data, was $7.8 trillion, while that of the United States was $14.29 trillion, 1.83 times that of China. See Central Intelligence Agency, *World Factbook, 2009.* The World Bank data show that China's 2007 GDP (PPP) was $7,083.5 billion, while that of the United States was $13,829.0 billion, 1.95 times that of China. World Bank, *World Development Report 2009: Reshaping Economic Geography*, Chinese ed. (Tsinghua University Press).

5. According to estimates by the CIA, China's exports made up 9.0 percent of the world's total in 2008, ranking second in the world, just after Germany (9.4 percent). CIA, *World Factbook, 2009.* According to the World Trade Organization report (WTO, *International Trade Statistics, 2009*), the export volume of China's service trade made up 3.9 percent of the world's total, ranking fifth in the world; China is likely to surpass France (4.2 percent), to rank fourth in the world.

6. William Overholt, *The Rise of China: How Economic Reform Is Creating a New Superpower* (New York: W. W. Norton, 1994); the Chinese scholars Justin Lin Yifu, Cai

Fang, and Li Zhou were the first to describe China's development as the China miracle. Justin Lin Yifu, Cai Fang, and Li Zhou, *China Miracle: Development Strategy and Economic Reform* (Shanghai: Shanghai People's Publishing House, 1994).

7. Timothy Garton Ash, "China, Russia, and the New World Disorder," *Los Angeles Times,* September 11, 2008 (www.latimes.com/news/opinion).

8. Howard W. French, "Letter from China: What if Beijing Is Right?" *International Herald Tribune*, November 2, 2007.

9. World Bank, *China: Socialist Economic Development*, vol. 3, *The Social Sectors: Population, Health, Nutrition, and Education* (1983).

10. World Bank, *China: Long-Term Development, Issues and Options* (Johns Hopkins University Press, 1985).

11. National Bureau of Statistics of China, *China Statistical Abstract 2010* (Beijing: China Statistics Press, 2010), p. 21.

12. John King Fairbank, *The Great Chinese Revolution 1800–1985* (New York: Harper-Collins, 1986; Chinese version, Beijing: World Affairs Press, 2000), p. 9.

13. Paul Kennedy, *The Rise and Fall of the Great Powers: Economic Change and Military Conflict from 1500 to 2000* (New York: Random House, 1987; Chinese version, Beijing: China International Culture Press, 2006), p. 554.

14. Richard Nixon, *1999: Victory without War* (New York: Simon and Schuster, 1988).

15. Maddison, "Statistics on World Population."

16. Angus Maddison, *Chinese Economic Performance in the Long Run, 960–2030 AD* (Paris: OECD, 2007), p. 13.

17. Charles Wolf and others, *Long-Term Economic and Military Trends 1994–2015: The United States and Asia* (Santa Monica, Calif.: Rand, 1995), p. 8.

18. World Bank, *China 2020 : Development Challenges in the New Century* (1997).

19. Ibid., p. 1.

20. Ibid., p. 21.

21. National Bureau of Statistics of China, *China Statistical Abstract 2010*, p. 24.

22. World Bank, *China 2020*, p. 36.

23. Asian Development Bank, *Emerging Asia: Changes and Challenges* (Manila: 1997).

24. Maddison, *Chinese Economic Performance in the Long Run.*

25. Jim O'Neill, "Building Better Global Economic BRICs," Global Economic Paper 66 (New York: Goldman Sachs, 2001).

26. Jim O'Neill, "Dreaming with BRICs: The Path to 2050," Global Economic Paper 99 (New York: Goldman Sachs, 2003).

27. Goldman Sachs, *Growth and Development: The Path to 2050* (New York: 2004); Goldman Sachs, *The World and the BRICs' Dream* (New York: 2006).

28. Goldman Sachs, "The Long-Term Outlook for the BRICs and N-11 Post Crisis," Global Economics Paper 192 (New York: 2009).

29. National Intelligence Council, *Mapping the Global Future* (Washington: 2004). Also see www.foia.cia.gov/2020/2020.

30. C. Fred Bergsten and others, *China: The Balance Sheet—What the World Needs to Know Now about the Emerging Superpower* (New York: Center for Strategic and International Studies and Peterson Institute for International Economics, 2006), pp. 3–4.

31. Albert Keidel, "China's Economic Rise: Fact and Fiction," July 2008 (http://carnegieendowment.org/files/pb61_keidel_final.pdf).

32. Maddison, *Chinese Economic Performance in the Long Run*, pp. 13–14.

33. Gregory C. Chow, *China's Economic Transition* (Oxford, U.K.: Blackwell, 2007), p. 9.

34. World Bank, *China 2020*.

35. Hu Angang, *Population and Development* (Hangzhou, Zhejiang: People's Publishing House, 1989).

36. Hu Angang, *Selected Works of Angang Hu: Ten Major Relationships in China's March toward the 21st Century* [Zhongguo Zouxiang Ershiyi Shiji de Shida Guanxi] (Harbin, Heilongjiang: Education Press, 1995), pp. 243–46.

37. Ibid., p. 341.

38. Ibid., pp. 122–23.

39. Deng Xiaoping, *Selected Works of Deng Xiaoping*, vol. 3 [Wenxuan] (Beijing: People's Publishing House, 1993), p. 374.

40. Hu, *Selected Works of Angang Hu*, pp. 243–46.

41. Hu Angang, "Knowledge and Development: China's New Catching-up Strategy" [Zhishi yu Fazhan: Ershiyi Shiji Xin Fazhan Zhanlue], in *Management World* [Guanli Shijie], no. 6 (1999): 7–24.

42. Hu Angang and Wang Yahua, *National Conditions and Development* (Beijing: Tsinghua University Press, 2005), p. 12.

43. Maddison, *Chinese Economic Performance in the Long Run*.

44. Hu Angang, *Roadmap of China's Rising* [Zhongguo Jueqi Zhilu] (Beijing: Peking University Press, 2007).

45. Terumasa Nakanishi, "How Can Broken Diplomacy Make Japan Win Out in the Great-Power Game?" *Seiron*, no. 8 (2008): 5.

46. Mao Zedong, *Selected Works of Mao Zedong,*" vol. 5, English ed. (Beijing: Foreign Languages Press, 1977), p. 312.

47. On February 2, 1975, when meeting the Gambian foreign minister Alieu Badara N'jie, Deng Xiaoping said, "Chairman Mao stressed that China will never seek hegemony, and that is the fundamental principle of our country. Some people say that it seems that China is a superpower. We really do not deserve it. We are not entitled to it. Why? We are very poor. We are a small country in economic level, and that is why we are not qualified to be a superpower. . . . Even in another twenty or thirty years, when China has 100 million tons of steel and have enough grain and to spare, shall we become a superpower? Chairman Mao called on us to educate well our future generations and never seek to be a superpower. Now we say that we belong to the third world. And even if in the future we become developed, we shall still belong to the third world. We should never be a superpower." CCP Central Committee Documentation Office, *Annals of Deng Xiaoping, 1975–1997*, vol. 1 (2004), p. 16.

48. Hu, *Population and Development.*

49. Human development level is measured by the human development index. A high HDI is between 0.8 and 0.9.

50. Joshua Cooper Ramo, *The Beijing Consensus* (London: Foreign Policy Center, 2004).

51. Toynbee holds that the growth of civilizations has four necessary conditions: cyclic movement of challenge and response, resolution of difficulties through methods that are internal and spiritual rather than external and material, a self-determining society (responding to internal challenges), and emergence of the dominant minority. In a word, the growth of civilization is driven by "creative minorities," who find solutions to the challenges, which others then follow by example, called mimesis. This general law governs the rise and growth of a civilization. Arnold Joseph Toynbee, *A Study of History* (Oxford University Press, 1934).

52. Hu Angang, Wang Shaoguang, and Zhou Jianming, *The Second Transition: Building a National System,* 2nd ed. (Beijing: Tsinghua University Press, 2008).

53. World Bank, *World Development Report 2010: Development and Climate Change,* p. 45.

Chapter Two

1. World Bank, *World Development Indicators 2009*, pp. 44–46.

2. *China Statistical Abstract 2010*, p. 177; *OECD Factbook 2009: Science and Technology.*

3. Alan Winters and Shahid Yusuf, eds., *Dancing with Giants: China, India, and Global Economy* (World Bank, 2007).

4. *China Statistical Abstract 2010*, pp. 24–25. But Angus Maddison estimates that from 1978 to 2006 China's GDP grew at an annual average 7.9 percent and per capita GDP grew at an annual average 6.7 percent, lower than the figures in the abstract. Angus Maddison, *Chinese Economic Performance in the Long Run, 960–2030 AD* (Paris: OECD, 2007), p. 21.

5. In 2007 China's GDP was $3.1260 trillion and that of Germany was $3.2073 trillion. World Bank, *World Development Indicators 2009*, p. 14. But my estimates are that China's GDP was $3.4073 trillion in 2008, more than that of Germany. According to the World Bank, U.S. GDP was 4.44 times that of China and Japanese GDP was 1.54 times that of China.

6. "China's Economy: Hello America," August 16, 2010 (www.economist.com/node/16834943).

7. Maddison, *Chinese Economic Performance in the Long Run*, p. 20.

8. Angus Maddison, "Statistics on World Population, GDP, and per Capita GDP, 1–2006 AD" (www.ggdc.net/maddison).

9. Ibid.

10. Maddison, *Chinese Economic Performance in the Long Run.*

11. Ibid. Maddison is the only consistent historical data source to show the long-term trend.

12. In 1956 Mao Zedong believed that China was beaten by imperialism because it was far behind. In 1979 Deng Xiaoping realized that the cause of China's weakness was its long-time policy of the closed door.

13. Francois Bourguignom and Christian Morrisso, "Inequality among World Citizens: 1820–1992," *American Economic Review* (September 2002): 727–44.

14. Angus Maddison, *The World Economy: A Millennial Perspective* (Paris: OECD, 2007), tables B-20, B-21.

15. Mao Zedong, "Report to the Second Plenary Session of the 7th Central Committee of the Chinese Communist Party," March 1949, in *Selected Works of Mao Zedong* (1967), p. 1320. According to the National Bureau of Statistics of China, the proportion of agriculture in GDP was as high as 50.5 percent in 1952, and secondary and tertiary industries made up 20.9 percent and 28.6 percent, respectively. *50 Years of New China: 1949–1999* (1999), p. 17.

16. National Bureau of Statistics of China, *New China 60 Years Statistics 2010*, p. 3.

17. Hu Angang, *China's Political Economic History, 1949–1976* (Tsinghua University Press, 2007), chap. 3.

18. Nicholas Crafts, "The Human Development Index, 1870–1999: Some Revised Estimates," *European Review of Economic History* 6, no. 3 (2002): 395–405.

19. Hu Angang, "From a Large Population Power to a Strong Power in Human Resources," *China Population Science*, no. 1 (2003): 1–10.

20. UN Development Program, *Human Development Report 2002*, p. 153.

21. World Bank Economic Study Group, *China: Socialist Economic Development, Main Report*, Chinese ed. (1982).

22. Comprehensive national power refers to the sum total of the power or strengths of a country's economy, military, science and technology, education, natural resources, and overall influence. I use eight types of indicators, which include twenty-three component indicators, to calculate comprehensive national power as a weighted index. This is a useful and comprehensive way to measure national power. See Hu Angang and Men Honghua, *International Comparison in Strategic Resources among China, the United States, India, and Russia: On China's Grand Strategy for Making the Country Strong and the People Rich*, Chinese ed. (Beijing: China Financial and Economic Press, 1982).

23. According to the PPP measures by the World Bank, China's per capita gross national income was only $340 in 1978 (current price), 52 percent of the average of the low-income bracket in the world, or 26 percent of the average of the middle-income bracket ($1,290). World Bank, *World Development Indicators 2002*.

24. World Bank, *China: Country Economic Memorandum—Sharing the Opportunities of Economic Globalization* (2003).

25. Angus Maddison, *Six Transformations in China, 960–2030* (www.ggdc.net/maddison)(2009).

26. Hu Angang, *The Political-Economic History of China, 1949–1976*, 2nd ed. (Tsinghua University Press, 2008), p. 217.

27. Li Xiannian, *Selected Works of Li Xiannian* (Beijing: People's Publishing House, 1981), pp. 423–24.

28. Bo Yibo, *Review of Major Policy Decisions and Events,* vol. 2 (Beijing: Central Party School Press, 1993), pp. 700–02.

29. Hu, *The Political-Economic History of China,* pp. 289–90.

30. Hu Angang, *Mao Zedong and the Great Cultural Revolution* (Hong Kong: Dafeng Publishing House, 2008).

31. Paul Krugman holds that China's input-driven growth, which was not driven by an increase in the efficiency with which those inputs are used, is inevitably limited. Paul Krugman, "The Myth of Asia's Miracle," *Foreign Affairs* 73, no. 6 (1994). Alwyn Young holds that the productivity performance of the nonagricultural economy during the reform period is respectable but not outstanding. Alwyn Young, "Gold into Base Metals: Productivity Growth in the People's Republic of China during the Reform Period," *Journal of Political Economy* 111, no.6 (2003).

32. Hu, *The Political-Economic History of China,* pp. 556–59.

33. Mao Zedong, "Where Do Correct Ideas Come From?" May 1963, in *Selected Readings from the Works of Mao Zedong* (Beijing: Foreign Languages Press, 1967), p. 405.

34. Angus Maddison, "Statistics on World Population, GDP, and per Capita GDP, 1–2008 AD" (www.ggdc.net/maddison).

35. Chen Yun, "Economic Situation and Experience and Lessons," December 16, 1980, in *Selected Works of Chen Yun,* vol. 3 (Beijing: People's Publishing House, 1995), p. 279.

36. Deng Xiaoping said, "What we do is an experiment. If errors occur, we must correct them immediately, seeing to it that no lesser errors turn into big blunders." Deng Xiaoping, "Reply to Queries from US Journalist Mike Wallace," September 2, 1986, in *Selected Works of Deng Xiaoping,* vol. 3 (Beijing: People's Publishing House, 1993), p. 174.

37. Their posts after 1956 were as members of the CCP Politburo Standing Committee and as vice premier.

38. Maddison, *Six Transformations in China.*

39. *China Statistical Abstract 2009,* p. 29.

40. World Bank, *World Development Indicators 2009,* pp. 232–34.

41. Zhou Xiaochuan, *Observations and Analysis of Savings Rate* (Beijing: People's Bank of China, 2009).

42. Ibid.

43. Gross capital formation refers to the fixed assets acquired minus disposals and the net value of inventory, thus including gross fixed capital formation and changes in inventories.

44. *China Statistical Abstract 2008,* p. 35.

45. World Bank, *World Development Indicators 2009,* p. 236.

46. Wang Xiaolu, Fan Gang, and Liu Peng, "China's Economic Growth Model Changeover and Sustainability," *Economic Studies,* no. 1 (2009): 4–16.

47. *China Statistical Abstract 2010,* pp. 13, 26, 44.

48. Ibid., p. 39.

49. National Bureau of Statistics of China, *Communiqué on Second National R&D Resources Survey (2010)*.

50. *China Statistical Abstract 2010*, p. 193.

51. I was directly involved in design and consultation for these two programs.

52. Wang Xiaolu and his coauthors forecast that China's economic growth would be 6.7–9.0 percent in 2008–20. Wang, Fan, and Liu, "*China's Economic Growth Model.*"

53. *China Statistical Yearbook 2010*, table 4-3.

54. Ibid.

55. I raised the point with Chinese leaders in 2001. My basic conclusion is that in the future the growth of labor and capital will have a very limited effect on China's economic growth and that the key factor lies in TFP. China's economic growth pattern will move from a "quantitative to qualitative model, that is, it will be centered round quality of growth, which should include not only average per capita income level but also the more equitable education and employment opportunities, greater gender equality and better health and nutrition, more sustainable natural environment, more just law administration and better legal system, more abundant cultural life and more effective social governance." Hu Angang, "Future Economic Growth of China Depends on TFP," *National Studies Report*, no. 46 (2001). World Bank experts hold that the driving force of China's high-speed growth is high TFP growth. In 2005–20 its growth may range from 3.5 to 4.0 percent. World Bank, *China Economic Quarterly*, June 2009.

56. According to OECD data, the number of urban families in Germany that have access to home computers was 69.9 percent in 2005 and that of Britain was 80.5 percent; that of the United States was 61.8 percent in 2003. *OECD Factbook 2007*. The number of urban families in China with access to home computers was 13.3 percent in 2001; the figure rose to 41.5 percent by 2005. *China Statistical Abstract 2007*, p. 119.

57. Urban spending on services in 2006 accounted for 52.5 percent of total home expenditure, while in 2000 the figure was 45.7 percent. *China Statistical Abstract 2007*, p. 118.

58. *China Statistical Yearbook 2010*, table 2-2.

59. One scholar holds that when per capita GDP (PPP) reaches $10,000, manufacturing would have reached its peak before slowing down or dropping. Barry Naughton, *Chinese Economy: Transitions and Growth* (MIT Press, 2007), p. 150.

60. Hu Angang and Lü Jie forecast that China's potential for sustainable economic growth will be around 9.5 percent. Hu Angang and Lv Jie, "Memo on China's Long-Term Sustainable Economic Development," manuscript, 2009.

61. China's automobile market was expected to overtake that of the United States. International Energy Agency, *World Energy Outlook* (2007).

62. This discussion is based on Hu Angang, "How Can China Narrow Its Relative GDP Gap with the United States?" *National Conditions Report*, no. 24 (July 2007), with added material.

63. According to the 2009 data collected by Angus Maddison, the U.S. GDP (PPP) was 1.17-fold that of China in 2008. Maddison, "Statistics on World Population." According

to CIA data, China's GDP (PPP) in 2008 was $7.80 trillion and that of the United States was $14.29 trillion, 1.83-fold that of China. Central Intelligence Agency, *World Factbook.* According to World Bank data, China's 2007 GDP (PPP) was $7.0835 trillion; that of the United States was $13.829 trillion, or 1.95-fold that of China. World Bank, *World Development Report 2000: Reshaping Economic Geography.*

64. Craig K. Elwell, Mare Labonte, and Wayne M. Morrison, *Is China a Threat to the U.S. Economy?* Congressional Research Service Report RL 33604 (2007).

65. Ibid.

66. The SNA is an accounting system for market economies; it uses such indicators as GNP, GDP, added value, end consumption (total consumption), total capital formation (total investment), and net export. Most countries in the world have adopted it. Simon Kuznets, in his report to the U.S. Congress in 1937, put forward the concept of GNP accounting. John Richard Nicholas (Dick) Stone, of Cambridge University and the Nobel laureate in economics in 1984, created the national accounts. Paul A. Samuelson and William D. Nordhaus, in their *Economics* (18th ed.), point out that "while the GDP and the rest of the national income accounts may seem to be arcane concepts, they are truly among the great inventions of the twentieth century. Much like a satellite in space can survey the weather across an entire continent, the GDP can give a whole picture of economic conditions. . . . The GDP . . . is like a lighthouse . . . to direct the economy." U.S. Secretary of Commerce William M. Daley said at a press conference that "as we searched for our greatest achievement, something the bright minds at Commerce created from scratch and that had the greatest impact on America, it was the invention of the national economic accounts— what we now call the gross domestic product, or GDP." Paul A. Samuelson and William D. Nordhaus, *Economics* (Boston: McGraw-Hill/Irwin, 2005); Steven Landefeld, "GDP: One of the Great Inventions of the 20th Century," *Survey of Current Business*, January 2000.

67. The uncertainty principle in quantum mechanics, formulated by Werner Heisenberg at Niels Bohr's Institute for Theoretical Physics at the University of Copenhagen, holds that it is not possible to measure the momentum of a particle and its position simultaneously to within less than an uncertainty limit.

68. Of the 500 top brands in the world in 2005, 249 were accredited to the United States, 46 to France, and 45 to Japan. Only 4, including CCTV, belonged to China. Of the first 100 most valued brands in the world, China has none.

69. Li Deshui, "New Changes on GDP and Its Structure in China" [Woguo GDP Zongliang he Jiegou Junyou Xin Bianhua], December 20, 2005, Xinhua News Agency (www.cctv.com/news/financial/inland/20051220/100695.shtml).

70. The number of people employed in the service trade in China grew an average annual rate of 3.3 percent in 2000–08.

71. World Bank, *World Development Indicators 2009*, pp. 208–09.

72. Angus Maddison and Harry Wu, "Measuring China's Economic Performance," *World Economics* 19, no. 2 (2008).

73. According to the forecast by Global Insight, China's GDP will grow during 2006–25 at an average annual rate of 7.1 percent and that of the United States at 3.0 percent;

by 2013 China will overtake the United States, to become the biggest economy in the world. China's GDP will be 10.1 percent higher by 2015 and 29.5 percent higher by 2020. According to forecasts by Economist Intelligence Unit, China would overtake the United States in GDP (PPP) by 2018. See Elwell, Labonte, and Morrison, "Is China a Threat to the U.S. Economy?" According to Carsten Holz of Hong Kong University of Science and Technology, China will overtake the United States in GDP during the period 2012–15 and by 2025 will become the biggest economic power in the world. Carsten A. Holz, "China's Economic Growth 1978–2025: What We Know Today about China's Economic Growth Tomorrow," 2006 (http://ssrn.com/abstract=756044).

74. *Selected Works of Mao Zedong,* vol. 5, English ed. (1977), p. 157.

75. Mao Zedong: "What is our objective in uniting with all the forces that can be united, inside and outside the party and at home and abroad? It is to build a great socialist country. A country like ours may and ought to be described as great. Our party is a great party, our people a great people, our revolution a great revolution, and our construction is great, too. Only one country on this globe has a population of 600 million, and that is China. In the past there were reasons for others to look down upon us. For we had little to contribute, steel output registered only several hundred thousand tons a year, and even that was in the hands of the Japanese. Under the despotic rule of Chiang Kai-shek's Kuomintang, which lasted twenty-two years, steel output was only some tens of thousands of tons a year. We still don't have much steel, but a promising start has been made. Output this year will be over 4 million tons. It will hit the 5 million mark next year, will exceed 10 million tons after the Second Five-Year Plan, and is likely to top 20 million after the Third. We must work hard to reach this target. There are about a hundred countries in the world, but only a few produce over 20 million tons of steel a year. Therefore, once built up, China will be a great socialist country and will radically transform the situation in which for over a century it was backward, despised, and wretched. Moreover, it will be able to catch up with the most powerful capitalist country in the world, the United States." *Selected Works of Mao Zedong,* pp. 314–15.

76. Maddison, "Statistics on World Population."

77. Angus Maddison, *Historical Statistics of the World Economy: 1–2006 AD* (www.ggdc.net/maddison).

78. Jeffrey D. Sachs, *The End of Poverty: Economic Possibilities for Our Time* (New York: Penguin, 2005).

79. Ibid.

Chapter Three

1. Mortality rate is typically expressed in yearly deaths per thousand individuals in a given region or country. See Miu Zhenpeng, "China's Population Problems in Semi-feudal Period," *Social and Economic Research in China,* no. 2 (1982).

2. UN Department of Economic and Social Affairs, *World Population Prospects: The 2008 Revision.*

3. Total fertility rate is the average number of children that would be born to a woman over her lifetime if she were to experience the exact current age-specific fertility rates through her lifetime and if she were to survive from birth through the end of her reproductive life. Central Intelligence Agency, *World Factbook, 2010.*

4. The demographic dividend is a rise in the rate of economic growth due to a rising share of working-age people in a population. World Bank, *World Development Report 2007: Development and the Next Generation,* pp. 35–36.

5. Angus Maddison, "Statistics on World Population, GDP, and per Capita GDP, 1–2003 AD" (www.ggdc.net/maddison).

6. On August 5, 1949, the U.S. State Department issued the white paper "U.S. Relations with China" and Dean Acheson's letter to Harry Truman. These can be found in Mao Zedong, *Selected Works of Mao Zedong,* vol. 4 (1961).

7. Ibid.

8. Hu Angang, "Mao Zedong and China's Population Growth," *Ming Pao Monthly,* no. 1 (1992).

9. *Renmin ribao* [People's Daily], editorial, July 9, 1978.

10. Deng Xiaoping, "Adhere to the Four Cardinal Principles," in *Selected Works of Deng Xiaoping,* vol. 2 (Beijing: People's Publishing House, 1979), pp. 163–64.

11. CCP Central Committee Documentation Office, "Open Letter by the CCP Central Committee to All Communist Party Members and Members of Communist Youth on the Control of Population Growth," September 25, 1980, in *Selected Major Documents since the Third Plenary Session of the 11th CCP Central Committee,* vol. 1 (Beijing: People's Publishing House, 1982), p. 538.

12. Hu Yaobang, *Report to the 12th National Party Congress of CCP,* September 1, 1982 (http://cpc.people.com.cn/GB/64162/64168/64565/65448/4526430.html).

13. Hu Angang and Zou Ping, *China's Population Development* (Beijing: China Science and Technology Press, 1991), p. 51.

14. UN Department of Economic and Social Affairs, *World Population Prospects: The 2008 Revision.*

15. In 1995 China's fertility rate was 1.87, lower than that of the United States, 1.98, and very close to the average of high-income countries, 1.71. World Bank, *World Development Indicators 2000 and Global Development Finance 2001.*

16. World Bank, *World Development Report 2007: Development and the Next Generation,* pp. 4–5.

17. Wang Feng and Andrew Mason, "The Demographic Factor in China's Transition," in *China's Great Economic Transformation,* edited by Loren Brandt and Thomas Rawski (Cambridge University Press, 2008), pp. 136–66.

18. UN Department of Economic and Social Affairs, *World Population Prospects: The 2008 Revision.*

19. *China Statistical Yearbook 2006,* tables 4-7, 4-16.

20. *China Statistical Yearbook 2009,* table 3-7.

21. World Bank, *World Development Indicators 2006.*

22. *China Statistical Abstract 2010*, p. 169.

23. *China Statistical Abstract 2009*, p. 38.

24. *China Statistical Abstract 2007*, p. 43.

25. UN Department of Economic and Social Affairs, *World Population Prospects: The 2008 Revision.*

26. Japanese National Conditions Society, *100-year Data of Japan*, 4th ed., p. 43; *Japanese National Conditions 2006–2007*, 64th ed., p. 50.

27. Japan's TFR in 2005 was 1.26; that of Hong Kong was 1.12; that of Singapore, 1.24; that of the Republic of Korea, 1.08; and that of Thailand, 1.80. Japan Ministry of Labor, *Japanese Population Developments 2007*, p. 12.

28. CCP Central Committee Documentation Office, "Open Letter of the CCP Central Committee to All Communist Party Members and Members of Communist Youth on the Control of Population Growth," September 25, 1980, in *Selected Major Documents since the Third Plenary Session of the 11th CCP Central Committee*, vol. 1, 1982, p. 538.

29. Zeng Yi, "Soft Landing with the Two-Children-Late-Birth Policy: Need and Feasibility" [Shilun Erhai Wanyu Zhengce de Biyaoxing he Kexingxing], *Social Sciences in China*, no. 2 (2006): 93–109.

30. UN statistics show that the slum population in India has reached 170 million, with Mumbai having the largest number of slums; out of Mumbai's total population of 11 million, about 55 percent constitutes the slum population. According to the statistics from Brazilian demographic research institutions, about 6.5 million Brazilians live in nearly 4,000 slums. Zhang Chuandu and others, "Slums Are the Roots of Turmoil in Developed Countries and Worries of Other Countries," *Global Times*, August 19, 2006.

31. Yang Xiaokai, "Notes on the Century's History of Chinese Economy (from late Qing to 1949)," *China Economic History Forum* (http://economy.guoxue.com/article.php/104).

32. National Bureau of Statistics of China, *50 Years of New China, 1949–1999* (1999), p. 31.

33. World Bank, *World Development Indicators 2010.*

34. UN Development Program, *Human Development Report 2002.*

35. Zhang Honglin and Song Shunfeng, "Population Movement and Urbanization" [Chengxiang Yimin yu Chengshihua], in *China's Urbanization: Empirical Analysis and Counter-Measure Studies* [Zhongguo de Chengshihua: Shizheng Fenxi yu Duice Yanjiu], edited by Cheng Yongjun and Chen Aimin (Xiamen University Press, 2002), p. 208.

36. Ministry of Agriculture, *Economy and Finance*, nos. 3, 4 (2003): 95.

37. World Bank, *World Development Indicators 2010.*

38. Eleventh Five-Year National Social and Economic Development Program of the People's Republic of China.

39. As a member of the expert panel of national programs, I participated in the preliminary background studies, policy formulation, and consultancy for the project.

40. Wang Lina, "China's Urbanization Process and Strategic Options," *Economic Study Reference*, no. 68 (2001): 43–45.

Chapter Four

1. Hu Angang, *SARS in Perspective: Health and Development* (Tsinghua University Press, 2003), p. 20.

2. National Bureau of Statistics of China, "Medicine and Health," in *50 Years of New China* (1999), esp. p. 86.

3. Ibid.

4. Minoru Kobayashi, *Key to China's Economic Development*, Chinese ed. (China Foreign Economic Relations and Trade Press, 1987).

5. UN Development Program, *Human Development Report 2005*. This report divides the HDI into three groups: low (0–0.50), middle (0.50–0.80), and high (0.80–1.0). The middle HDI may be subdivided into lower middle (0.50–0.75) and upper middle (0.75–0.80).

6. UN Development Program, *Human Development Report 2002*.

7. Age-standardized mortality rates are used to compare different locales' mortality rates without being skewed by the difference in age distributions from place to place. See Charles Wolf Jr. and others, *Fault Lines in China's Economic Terrain*, trans. Xu Jing (Beijing: Xinhua Publishing House, 2005), p. 46.

8. Ministry of Health of the People's Republic of China, *Third National Health Survey 2003; Fourth National Health Survey 2009*.

9. David Bloom and David Canning, "Schooling, Health, and Economic Growth: Reconciling the Micro and Macro Evidence," working paper (Santa Monica, Calif.: Rand, 2005).

10. Jeffrey D. Sachs and Wing Thye Woo, "Understanding China's Economic Performance," *Journal of Policy Reform* 4, no. 1 (2000): 1–50.

11. Amartya Sen, *Development as Freedom*, Chinese version, trans. Ren Ji and Yu Zhen (Chinese People's University Press, 2002), p. 34.

12. Hu Linlin and I constructed a production function, with health condition as an input of human capital, and then used statistical data to estimate each input factor's contribution to economic growth. See Hu Linlin, "Health and China's Economic Growth: Theoretical Framework and Empirical Analysis," Ph.D. dissertation, Tsinghua University. Hu Linlin worked under my tutorship.

13. Peter Nolan and John Sender, "Death Rates, Life Expectancy, and China's Economic Reforms: A Critique of A. K. Sen," *World Development* 20, no. 9 (1992): 1279–303; Amartya Sen, "Life and Death in China: A Reply," *World Development* 20, no. 9 (1992): 1305–12.

14. Shaoguang Wang, "Background Report," in UN Development Program, *Human Development Report 2005*.

15. Hu Angang and Linlin Hu, "Macroeconomy and Health in China," *Reform*, no. 2 (2003): 5–13.

16. Hu Angang, "Health Insecurity: The Biggest Challenge to Human Security in China," in Hu, *Economic and Social Transformation in China: Challenges and Opportunities* (London: Routledge, 2007), pp. 152–66.

17. The patient group is made up of self-reported patients. Specifically, it includes those who did not feel well and visited a doctor, those who did not feel well and self-medicated,

and those who did not feel well and stayed home for more than one day as a result. One of the cases is regarded as having contracted a disease. Ibid., p. 154.

18. Chronic diseases include such infectious diseases as TB and such noninfectious diseases as coronary artery heart disease and high blood pressure. To be included in the count, the cases had either to have been diagnosed at least six months before the survey or to have recurred at least six months before the survey and were being treated.

19. Hu, *Economic and Social Transformation in China,* p. 155.

20. Ministry of Health of the People's Republic of China, *Third National Health Survey.*

21. Ministry of Health, *Health Statistical Yearbook 2004.*

22. Ministry of Health, *Third National Health Survey.*

23. Ministry of Health, *Health Statistical Yearbook 2009,* table 7-1.

24. UN Development Program, *Human Development Report 2009/2010,* p. 131.

25. World Bank, *World Development Report 2002.*

26. *China Statistical Yearbook 2004,* p. 560.

27. Alcoholics are those who drink at least three times a week; a drinking habit is formed if one drinks at least 156 times a year.

28. Calculation based on *China Statistical Yearbook 2004,* pp. 95, 366, 389.

29. Xinhua News Agency, "300 Million People Do Not Have Access to Clean Water," November 28, 2004.

30. Ministry of Water Resources of the People's Republic of China, *Research Report on Rural Water Use Survey at the County Level 2005.*

31. World Health Organization, *Global Health Observatory 2010.*

32. *China Statistical Yearbook 2010,* tables 21-43, 21-44.

33. Ministry of Health of the People's Republic of China, UNAIDS, and WHO, *AIDS Epidemic Prevention Work Progress in China* (2005).

34. UNAIDS, *2008 Report on the Global AIDS Epidemic,* p. 219.

35. Wolf and others, *Fault Lines in China's Economic Terrain.*

36. Wang Shaoguang, "Crisis and Turnaround of China's Public Health," *Comparative Studies,* no. 7 (2003).

37. Drinking water not up to safety standards refers to water with a high content of fluorine and arson, bitter and salty water, polluted water and water in snail-fever infested areas, and water with microorganisms exceeding set standards. Ma Kai, *Reader in Aid of Understanding of the 11th Five-Year National Economic and Social Development Program of the People's Republic of China* (Beijing: Science and Technology Press, 2006), p. 517.

38. The percentage for middle HDI countries in 2000 was 6.9 percent. UN Development Program, *Human Development Report 2003,* p. 103.

39. Jiang Zemin, "Report to the 16th CCP National Congress," in *Selected Major Documents since the 16th CCP National Congress,* p. 14.

40. UN Development Program, *Human Development Report 2008,* p. 208.

41. Angus Maddison, *The World Economy: A Millennial Perspective* (Paris: OECD, 2007).

42. UN Development Program, *Human Development Report 2005,* p. 22.

43. Ibid.; UN Development Program, *Human Development Report 1995*, p. 20; UN Development Program, *Human Development Report 2010*, p. 20.

Chapter Five

1. Easterlin Kichard, "Why Isn't the Whole World Developed?" *Journal of Economic History* 41, no. 1 (1981): 1–17.

2. *China Education Yearbook 1949–1982*, p. 168.

3. Other totals: the United States, 11.27 years; France, 9.58 years; Germany, 10.40 years; Japan, 9.11 years. Angus Maddison, *Chinese Economic Performance in the Long Run, 960–2030 AD* (Paris: OECD, 2007).

4. Total human capital refers to the product of the number of people in the fifteen-to-sixty-four-year-old bracket multiplied by the average years of education received. Person-year is defined as one year or fraction thereof of education by an education institute.

5. *China Statistical Abstract 2010*, p. 174.

6. Hu Angang, "From a Big Power in Population to a Big Power in Human Capital: 1980–2000," *China Demography*, no. 5 (2002): 1–10.

7. For detailed analysis, see Hu Angang, "Historical Appraisal of the Mao Era," in *History of China's Political Economy, 1949–1976* (Tsinghua University Press, 2007).

8. National Bureau of Statistics of China, *New China 60 Years Statistics 2010*, table 1-7-1.

9. *China Education Yearbook 1949–1982*, p. 922.

10. *China Statistical Abstract 2009*, pp. 185, 190.

11. Wang Xiaolu, Fan Gang, and Liu Peng, "China's Economic Growth Model Changeover and Sustainability," *Economic Studies*, no. 1 (2009): 4–16; Hu, "From a Big Power in Population to a Big Power in Human Capital."

12. The first third-party independent evaluation of the implementation of the Tenth Five-Year Plan points out that China failed to realize this crucial indictor. See Hu Angang, Wang Yahua, and Yan Yilong, "Evaluation Report on Implementation of Tenth Five-year Plan," *Review of Economic Research*, no. 2 (2006): 40–55.

13. Voice of America, "Education Report Foreign Student Series: Financial Aid," February 4, 2009.

14. Ministry of Education of the People's Republic of China, news briefing, March 25, 2009.

15. Ministry of Education, *Yearbook of International Cooperation* (2009) (www.moe. gov.cn/publicfiles/business/htmlfiles/moe/s3124/201002/82571.html).

16. Voice of America, "Education Report Foreign Student Series."

17. O'Malley, "US Share of Foreign Students Drops."

18. Ministry of Education, news briefing.

19. OECD, *Education at a Glance 2008* (Paris), pp. 59, 68.

20. Hu, "From a Big Power in Population to a Big Power in Human Capital."

21. China Education and Human Resources Development Project, *From a Big Power in Population to a Strong HR Power* (Beijing: Higher Education Press, 2003). In October 2005 Party General Secretary Hu Jintao, at the fifth plenary session of the Sixteenth Central Committee, called for "efforts to stimulate the capacity building in HR and improve the general qualities of laborers so as to turn China from a big power in population into a strong HR power." Based on background report by Hu Angang, March 2009.

22. UN Department of Economic and Social Affairs, *World Population Prospects: The 2008 Revision.*

23. Wang Xiaolu, Fan Gang, and Liu Peng, "China's Economic Growth Model Changeover and Sustainability," *Economic Studies*, no. 1 (2009): 4–16.

24. Spending on basic research by institutions of higher learning in 2007 made up 50 percent of the total spending in the country. The number of papers published internationally by the first authors accounted for over 70 percent of the national total.

25. World Bank, *World Development Indicators 2009*, pp. 14–16, 80–82.

26. In 1997 China revised the fees for students from abroad who cover their own expenses: undergraduates in liberal arts, 14,000–26,000 yuan a year; master's degree candidates, 18,000–30,000 yuan a year; Ph.D candidates, 22,000–34,000 yuan a year; short-term students, 3,000–4,800 yuan a month or 8,000–10,000 yuan for three months; science and engineering students, 10–30 percent more than undergraduates in liberal arts; medicine, arts, and sports students, 50–100 percent more than undergraduates in liberal arts.

Chapter Six

1. This golden age started actually in 1978, but it was not recognized until recently.

2. The World Bank's *East Asian Miracle: Economic Growth and Public Policy,* published in 1993, does not include China in the "miracle." However, the World Bank's *An East Asian Renaissance: Ideas for Economic Growth*, published in 2007, calls China the biggest success story in development, citing China's GDP (accounting for half of the East Asian total) and its exports (accounting for a third of the East Asian total). China has a special place in the story of East Asia because of its absolute size, its unusual openness for a continental economy, and its orientation toward the region. China is now the world's third-largest trader and is the largest trader in East Asia, having overtaken Japan in 2004. Indermit Gill and Homi Kharas, *An East Asian Renaissance: Ideas for Economic Growth* (World Bank, 2007).

3. This section expands upon Hu Angang and Xiong Yizhi, "Quantitative Evaluation of China's Power in Science and Technology, 1980–2004," *Tsinghua University Gazette* (Philosophy and Social Sciences ed.), no. 2 (2008).

4. World Bank, *World Development Report: Knowledge for Development 1998.*

5. Hu Angang, *Knowledge and Development* (Beijing University Press, 2001).

6. *China Statistical Abstract 2010*, p. 177; OECD, *Main Science and Technology Indicators 2010*, 2nd ed.

7. Ministry of Science and Technology of the People's Republic of China, *Collection of Statistical Data in Science and Technology 2010;* National Science Foundation, *US Science and Engineering Indicators 2010*.

8. Jiang Zemin, *Sixteenth National Party Congress Report*.

9. Hu Angang and Men Honghua, "International Comparison of Composite National Power among China, USA, Japan, Russia, and India, 1980–1998," *National Conditions Report*, no. 10 (2002).

10. Jong-Hak Eun, "Assessing the Science and Technology Power of China," paper prepared for the conference Assessing the Power of China: Political, Economic, and Social Dimensions, Seoul, May 30, 2007.

11. Hu Angang and Xiong Yizhi, "The Quantitative Evaluation of China's Scientific and Technological Strength," *Journal of Tsinghua University* 2 (2008).

12. R&D refers to systematic and creative work that increases the total amount of knowledge (including human, cultural, and social knowledge) and that applies this knowledge innovatively.

13. The OECD also uses the PPP method in calculating R&D expenditure. Martin Schaaper says that the size of its R&D expenditure has made China a global R&D player, ranking only behind the United States and Japan in purchasing power parity (PPP) terms. Martin Schaaper, "Measuring China's Innovation System: National Specificities and International Comparisons," STI Working Paper 2009/1 (OECD, 2009).

14. Manuel Trajtenberg has discovered that the more investment a country makes in R&D, the greater the potential capacity a country has and the more innovations it generates. Manuel Trajtenberg, "Product Innovations, Price Indices and the Measurement of Economic Performance," Working Paper 3261 (Cambridge, Mass.: National Bureau of Economic Research, 2005).

15. *Collection of Statistical Data in Science and Technology 2010*.

16. *Statistics and Analysis of Chinese Papers in S&T 2009*.

17. Eun, "Assessing the Science and Technology Power of China"; P. Zhuo and L. Leydesdorff, "The Emergence of China as a Leading Nation in Science," *Research Policy* 35, no.1 (2006): 83–104.

18. *Collection of Statistical Data in Science and Technology 2010*.

19. *Statistics and Analysis of Chinese Papers in S&T 2009*.

20. *Collection of Statistical Data in Science and Technology 2009*.

21. China Science and Technology Information Institute, *Statistical Data of Chinese S&T Papers, 2009*.

22. "The National Medium- and Long-Term Program for Science and Technology Development (2006–2020): An Outline."

23. World Intellectual Property Organization, "Intellectual Property Statistics 2010" (www.wipo.int/ipstats/en/).

24. The number of invention patents granted by China in 2007 was 8.9 percent of the world's total (just after Japan, the United States, and South Korea), to rank fourth in the world. World Intellectual Property Organization, *World Intellectual Property Indicators 2009* (Geneva), p. 17.

25. Francis Gurry, "Message to the 2009 World Intellectual Property Right Day," March 30 (Geneva: World Intellectual Property Organization, 2009).

26. *China Statistical Abstract 2007*, pp. 19, 202; *China Statistical Abstract 2010*, p. 177; author's calculations.

27. *China Statistical Abstract 2010*, pp. 19, 70.

28. Zhang Chunlin and others, *China: Promoting Enterprise-Based Innovation* (World Bank, 2009).

29. According to OECD estimates, China's R&D expenditure in 2006 was $130 billion, overtaking Japan to become the world's second-largest spender in R&D, just after the United States ($330 billion). "China Cranks up Investment in R&D," *Financial Management* (London), February 2007, p. 4.

30. OECD, *Main Science and Technology Indicators 2008*, 2nd ed.

31. World Bank, *World Development Indicators 2006*.

32. *China Statistical Abstract 2010*, pp. 141, 112, 118.

33. Ibid., p. 161; U.S. Federal Communications Commission, "Internet Access Services: Status as of June 30, 2009," p. 6.

34. Website of *Point Topic*, October 3, 2008 (http://point-topic.com/).

35. *Selected Works of Mao Zedong*, vol. 5, p. 306.

36. The *Fengqing* was built by the Shanghai Jiangnan Shipyard, but the Ministry of Communication was unhappy with its standards, even as other State Council bodies budgeted money for buying ships from abroad (supported by Zhou). These concrete disputes were subsumed by larger ideological themes, with adherence to Mao's self-reliance injunctions as the correct line in contrast to buying from abroad. CCP Central Committee Documentation Office, *Biography of Mao Zedong 1949–1976*, vol. 2, pp. 1701–02.

37. Deng Xiaoping, "Respect Knowledge, Respect Talents," in *Selected Works of Deng Xiaoping*, vol. 2, p. 40.

38. Deng Xiaoping said, "Now we must learn what is advanced from foreign countries, practicing 'take-overism.' Japan has developed fast in that it resorted to 'take-overism.'" CCP Central Committee Documentation Office, *Annals of Deng Xiaoping, 1975–1977*, vol. 1 (2004), p. 236.

39. Hu Angang, *China's Political and Economic History, 1949–1976* (Tsinghua University Press, 2007).

40. OECD countries (including, in Asia, Japan and South Korea) contributed 90.8 percent of total invention patents. World Bank, *World Development Indicators 2006*.

41. Li Ping's analysis of the way import trade stimulates technical innovation includes two steps. First, international merchandise trade brings in motives of imitation, thus generating the overflow effect and enhancing the technical level. Second, the competitiveness

of imported goods affects the market shares of domestic manufacturers and thus indirectly stimulates technical innovation. Li Ping, "On the Relations between International Trade and Technical Innovation," *World Economic Studies*, no. 5 (2002).

42. *Statistical Bulletin on the 2006 National Economic and Social Development of the People's Republic of China*, February 28, 2007.

43. World Bank, *World Development Indicators 2006;* author's calculation.

44. According to the World Bank, between 1990 and 2003 internal trade in East Asia rose from 42 percent to 54 percent, while in North America the proportion rose from 37 percent to 45 percent. Gill and Kharas, *An East Asian Renaissance*.

45. Dongguan City in Guangdong Province reported a GDP growth of 144 times between 1980 and 2005, averaging 22 percent annually. It is regarded by the World Bank as "a successful case that integrates the three major theories of new economic growth, new international trade, and new economic geography." Ibid., p. 12.

46. The inflow of FDI can stimulate the independent R&D of internal capital enterprises. Wang Hongling, Li Daokui, and Feng Junxin, "FDI and Independent R&D: Data-Based Empirical Studies," *Economic Studies*, no. 6 (2006).

47. UN Conference on Trade and Development website; *China Statistical Abstract 2009*, p. 179.

48. *China Statistical Abstract 2010,* p. 177.

49. China Patent Law provides that the validity terms of international patents is twenty years.

50. National Bureau of Statistics, *2009 Statistics Bulletin of the National Economic and Social Development of China*.

51. *China Statistical Abstract 2010,* p. 177.

52. China Science and Technology Information Institute, *Statistical Data of Chinese S&T Papers, 2009*, pp. 8–9.

53. The business sector has become the largest R&D performer in terms of inputs, outputs, and patent applications, and it finances the largest share of its own R&D activities. Schaaper, "Measuring China's Innovation System." The government document adopted in 1995 ("Decisions of the CCP Central Committee and the State Council on Accelerating the Pace of S&T Progress") issued the call to "display to the full the market mechanism to stimulate the development of science and technology." In 1999 the official document (*Notice on Promotion of Scientific and Technological Achievements*) made it even clearer that "the commercialization and industrialization of technical innovations and high and new technologies must be based on market demand, social demand, and demand by state security and must make enterprises the main players in innovation and bring into full play the basic functions of the market in the allocation of science and technology resources and encourage most scientific and technical forces to enter the market for innovation and pioneering new undertakings."

54. Ministry of Science and Technology of the People's Republic of China, *S&T Statistics Data Book 2009*, table 20-38.

55. P. Mueller, "Exploring the Knowledge Filter: How Entrepreneurship and University-Industry Relationships Drive Economic Growth," *Research Policy* 35 (2006): 1499–508.

56. Zhang and others, *China.*

57. *China Statistical Yearbook 2009*, table 20-54.

58. Ibid., table 20-62.

59. Ministry of Science and Technology of the People's Republic of China, "Statistical Analysis Report on the National Technology Market in 2009," July 2010.

60. According to the U.S. National Science Foundation's analytical framework regarding national competitiveness in science and technology, national orientation should include a national strategy for promoting technology development, a social impact conducive to technical progress, an entrepreneur spirit, and investment risks. Based on this framework, the indicator of China's competitive power in science and technology has always been higher than other indicators (social and economic base, technical infrastructure, and production capacity). This shows that a powerful national orientation is the advantage enjoyed in China's science and technology development.

61. Deng Xiaoping, "Speech at the Opening of the National Science and Technology Conference," March 18, 1978, in *Selected Works of Deng Xiaoping*, vol. 2, p. 86.

62. "Decisions of the CCP Central Committee on the Reform of the Science and Technology System," in *Selected Major Documents since the 12th National Party Congress*, vol. 2, p. 671.

63. Jiang Zemin, "Implement the Strategy of Invigorating the Nation by Developing Science and Education," in *Selected Works of Jiang Zemin*, vol. 1, p. 428.

64. The goals set for 2020 by the program are the following: raising the proportion of R&D input in GDP to over 2.5 percent, raising the contributions to technical progress to over 60 percent, lowering dependence on foreign technology to less than 30 percent, raising the number of Chinese patents, and bringing the number of cited international papers into the first five in the world.

65. *The Economist*, April 25, 2009.

66. *China Statistical Yearbook 2009*, table 7-6.

67. K. M. Murphy, A. Shleifer, and R. W. Vishny, "Industrialization and the Big Push," *Journal of Political Economy* 97, no. 5 (1989): 1003–25.

68. On April 29, 2007, U.S. National Academy of Sciences President Ralph Cicerone said that Yuan Longping made outstanding contributions to the world food security by creating hybrid rice and that the increase in food solved the food problems of 35 million people in the world. See Zeng Penghui, "Yuan Longping Is Elected as Fellow of US Academy of Science" (www.gmw.cn/content/2007-05/22/content_610625.htm), May 22, 2007.

69. Fan Shenggen, "Infrastructure and Pro-Poor Growth," paper prepared for the World Bank Transport Forum, March 29, 2006.

70. Hu Jintao, speech, National Science and Technology Conference, 2006.

71. The general goals of science and technology development for 2020 are to significantly enhance the country's independent innovation capacity and the capacity of science

and technology to stimulate economic and social development, ensure national security, and build a well-off society. It is hoped that by the middle of this century both basic sciences and frontier technologies will advance China into the ranks of innovative countries. "The National Medium- and Long-Term Program for Science and Technology Development (2006–2020)."

72. China became first in the world in terms of human resources in science and technology in 2007, when it had 42 million scientists and technicians, including about 18 million with at least a university education. *Collection of Statistical Data in Science and Technology 2009.* Ma Kai, ed., *Supplementary Reading Concerning the 11th Five-Year Program for the National Economic and Social Development of the People's Republic of China* (Beijing: Beijing Science and Technology Press, 2005), p. 342.

73. Zhang and others, *China.*

74. "The National Medium- and Long-Term Program for Science and Technology Development (2006–2020)." In 2007 China was fourth in the world in terms of invention patent grants, accounting for 9.6 percent of the world total; in number of papers published internationally it ranked second; in number of papers cited internationally, it ranked tenth. *Collection of Statistical Data in Science and Technology 2009.*

75. China's high-tech exports in 2007 totaled $373.8 billion, which was 29 percent of total merchandise export and 1.18 times U.S. high-tech exports in 2006 ($318.0 billion). *Collection of Statistical Data in Science and Technology 2009.*

76. In 2009 more than 700,000 highly skilled residents in OECD countries were Chinese-born; 57 percent of them were living in the United States. Schaaper, "Measuring China's Innovation System."

77. Mike Wallace and Bill Adlor, preface, *The Way We Will Be 50 Years from Today* (Beijing: China Youth Press, 2009).

78. Gurry, "Message to the 2009 World Intellectual Property Right Day."

Chapter Seven

1. The 2008 Nobel Prize winner for economics, Paul Robin Krugman, said in his lecture at Shanghai Jiaotong University on May 12, 2009: "From a long-term perspective, environmental policy will dominate all other policies, as the environmental problem is more crucial than the financial system and international trade. Such a trend may not be seen next year, but in ten years, fifteen years, especially with climate change, it will become the center of all social activities and the economy." Jin Ji, *Xinmin Zhoukan* [Xinmin Weekly], May 25, 2009.

2. For the full text of the report, see the IPCC official website, www.ipcc.ch.

3. Take the main greenhouse gas, CO_2, for instance: its temperature-boosting effect accounts for about 60 percent of the total effect of GHGs. It may stay in the air for 30–3,000 years. Qin Dahe, "Challenges of Climate Change to China's Sustainable Economic and Social Development," *Guangdong Study Forum*, no. 43 (2007).

4. UNDP, *Human Development Report 2007/2008—Fighting Climate Change: Human Solidarity in a Divided World.*

5. Susan Woodward, "Tropical Broadleaf Evergreen Forest: The Rainforest," 1997 (www.radford.edu/~swoodwar/CLASSES/GEOG235/biomes/rainforest/rainfrst.html).

6. *China Statistical Yearbook 2007*, p. 4.

7. Working Group III, "Mitigation of Climate Change," in *IPCC Fourth Assessment Report.*

8. National Development and Reform Commission, *China's National Scheme of Coping with Climate Change 2007.*

9. One estimate is that 100 million people will be "directly at risk from coastal flooding," particularly low-lying countries in the Pacific, most of Bangladesh, and big cities such as Shanghai, Hamburg, Bangkok, Jakarta, Mumbai, Manila, Buenos Aires, London, and Venice. A. Dupont and G. Pearman, "Heating up the Planet: Climate Change and Security," Paper 12 (Lowy Institute for International Policy, 2006).

10. *China's National Scheme of Coping with Climate Change 2007.*

11. For more discussion, see Xiao Guoliang, *Imperial Power and Chinese Socio-Economy* (Beijing: Xinhua Publishing House, 1991).

12. Shamanthy Ganeshan and Wayne Diamond, "Forecasting the Numbers of People Affected Annually by Natural Disasters up to 2015," 2009 (www.oxfam.org/sites/www.oxfam.org/files/forecasting-disasters-2015.pdf).

13. Author's calculations, based on OFDA/CRED international disaster database.

14. National Bureau of Statistics of China, *2008 Annual Report on the National Economy and Social Development.*

15. Green Peace and Oxfam, "Climate Change and Poverty: A Case Study of China," 2009 (www.oxfam.org.cn/userfiles/report/CC.poverty.report.pdf).

16. According to estimates by the International Energy Agency, China consumed 1.563 billion tons of standard coal in 2005, accounting for 37.6 percent of the world total, only 3.3 percent less than that consumed by all the OECD countries combined (1.615 billion tons). We estimate that raw coal consumption in 2006 could top the OECD total. China produced 1.636 billion tons of standard coal in 2005, or 39.4 percent of the world total (4.154 billion tons). IEA, *World Energy Outlook 2007*. China's SO_2 emission was 26 million tons in 2005; it may increase to 31 million tons by 2015, according to the IEA. China emitted 5.1 billion tons of CO_2 in 2005, second only to the U.S.'s 5.8 billion tons. According to the IEA, by 2015 China's CO_2 emissions could reach 860 million tons (overtaking the U.S.'s 640 million tons—and ranking first in the world). IEA, *World Energy Outlook 2007.*

17. International Energy Agency, "China Overtakes the United States to Become World's Largest Energy Consumer," 2010 (www.iea.org/index_info.asp?id=1479).

18. Climate change has become an important subject of discussion in international cooperation as well as in Sino-U.S. strategic cooperation. According to the Obama administration, strategic cooperation can be broken into four subjects: climate change, energy, global financial crisis, and armament control. The first two are the most important. *China Youth News* (Washington), July 22, 2009.

19. IEA, *World Energy Outlook 2002*; IEA, *Key World Energy Statistics, 2010*.

20. Tao Wang and Jim Watson, "Who Owns China's Carbon Emissions?" Briefing 23 (Tyndall Centre for Climate Change Research, 2007). Also see http://tyndall.webapp1. uea.ac.uk/publications/briefing_notes/bn23.pdf.

21. Li Liping, Ren Yong, and Tian Chunxiu, "China's Carbon Emission Responsibilities from the International Trade Perspective," *Environmental Protection*, no. 6 (2008).

22. Chatham House, "Changing Climates: Interdependencies on Energy and Climate Security for China and Europe," 2007, p. 12.

23. *China Statistical Yearbook 2009*, table 6-2.

24. Huang Haifeng and Gao Nongnong, "Adjust the Industrial Structure and Open up a New Path to Environmental Protection," *Environmental Protection*, no. 6 (2009).

25. Zheng Wang, "Mixed Policies Favor Climate Protection," *Science Daily*, April 25, 2007.

26. State Forestry Administration, *Bulletin on China's Forestry and Ecological Construction* (2008).

27. According to data provided by the Union of Concerned Scientists, China's GHG emissions reached 6 billion tons in 2006, exceeding U.S. emissions (5.9 billion tons). Per capita emissions were 4.6 tons, while U.S. emissions were 19.8 tons, 4.3 times more than China's.

28. In 1990–2005 China's CO_2 emissions rose from 2.244 billion tons to 5.101 billion tons, averaging an annual growth of 5.6 percent. Its share of world CO_2 emissions rose from 10 percent to 19 percent. According to an IEA estimate, by 2015 its emissions could increase to 8.632 billion tons, 35 percent higher than the United States figure, or averaging an annual growth of 5.4 percent. Even if its CO_2 emission growth drops in 2015–20, it could still reach 8.9 billion tons, or 52 percent of newly added emissions in the world.

29. The U.S. energy secretary, Steven Chu, says that leadership by the United States and China (in emissions reduction) will determine, to a great extent, the destiny of the world. *Global Times*, July 16, 2009.

30. The power consumption of industry in 2008 was 2,549.5 billion kilowatt hours, or 73.9 percent of the national total. Zhang Guobao, *China Energy Development Report 2009* (Beijing: Economics Press, 2009), p. 94.

31. Author's calculation, based on *China Statistical Abstract 2009*, pp. 21, 146.

32. *China Environment Statistical Yearbook 2007*, pp. 4–5.

33. Energy-gobbling industries are industries whose energy consumption accounts for over 1.5 times the percentage of their industrial output. The added value of the listed industries accounted for a fifth of industrial added value, but their energy consumption was two-thirds of the total. IEA, *World Energy Outlook 2007*.

34. According to quantitative estimates by the Tianze Economic Institute, the direct external losses caused by the use of coal is about 1.7903 trillion yuan if calculated by the 2007 coal price and output level, accounting for 7.3 percent of the year's GDP. Tianze Economic Institute, *Coal Cost and Price Formation and the Internalization of External Cost* (Beijing: 2009).

35. In 2005 the State Council set the goal of resolving the small coal pits problem in about three years and formulated a three-year plan to correct or shut down small coal

mines. But the number of small coal pits shut down in 2008 was only 1,054, or 7.5 percent. Coal mines with a backward production capacity of over 40 million tons a year that were rejected accounted for 1.6 percent.

36. The thermal power plants with soot desulfurization devices in operation in 2002–07 exceeded 270 million kilowatts, or 50 percent of the national total thermal power output, thus greatly reducing the discharge of soot and SO_2. Zhang, *China Energy Development Report 2009*, p. 35.

37. Jiang Kezhun lists eight major technologies: renewable energy; advanced nuclear power; fuel cells; advanced clean coal, carbon capture, and storage; advanced gas-fired turbines; nonregular natural gas and crude oil production; synthetic fuel; super-low energy consumption and zero-emission transport. Jiang Kezhun, "China's Energy Demand and Greenhouse Gasses Emissions," in *World Climate Diplomacy and China's Counter-Measures*, edited by Yang Jiemian (Beijing: Current Affairs Publishing House, 2009).

38. International Energy Agency, *World Energy Outlook 2007: China and India Insights*.

39. Zhang, *China Energy Development Report 2009*, pp. 283–84.

40. Quoted in *Global Times*, July 16, 2009.

41. The proportion of China's net energy export in the total energy consumption was 18 percent in 2001 and 28 percent in 2004. CO_2 emissions were about 1.1 billion tons, accounting for 23 percent of the total. Huang Haifeng and Gao Nongnong, "Restructuring the Industry and Pioneering a New Path to Environmental Protection," *Environmental Protection*, no. 6 (2009).

42. By taking voluntary actions to reduce GHG emissions, China as a developing country acts according to its own reality and national conditions and is not restricted or limited by such international conventions as the Kyoto Protocol.

43. When Deng Xiaoping visited Japan during October 22–29, 1978, he answered a reporter's question about the Diaoyu (Senkaku) island territorial disputes.

Chapter Eight

1. Zhou Chengcheng and others have published a study on how to integrate the objectives for building both a well-off society and the millennium development goals into China's development plan. UN Development Program, "Building a Well-Off Society in China: A National Development Program that Integrates MDGs," 2004, Project RR/03/M02/A/PJ/99.

2. Jiang Zemin, "Build a Well-Off Society in an All-Round Way and Break New Ground in Building Socialism with Chinese Characteristics," in CCP Central Committee Documentation Office, *Selected Major Documents since the Sixteenth Party Congress*, pt. 1 (2005).

3. "Decisions of the Central Committee of the Communist Party of China on Major Issues Concerning the Construction of a Socialist Harmonious Society," adopted at the sixth plenary session of the Sixteenth Central Committee, October 11, 2006.

4. UN Development Program, *Human Development Report 2003*.

5. According to Zeng Peiyan, speaking at the Sixteenth Party Congress in 2002, China will basically complete industrialization in the first two decades of the century, which is necessary for China to modernize by the middle of the century. The major tasks for completing industrialization are to quadruple the 2000 GDP by 2020, with per capita GDP increasing significantly, urbanization level raising markedly, the proportion of agricultural labor dropping sharply, and the proportion of the service industry and industrial technology rising by a big margin. Per capita GDP in 2020 could reach more than $3,000, by and large the same as middle-income countries. By 2020 urbanization could reach over 50 percent; the proportion of agricultural labor could drop from 50 percent in 2000 to about 30 percent in 2020. Zeng Peiyan, *Reader for Understanding the Sixteenth Party Congress Report* (Beijing: People's Publishing House, 2002), p. 79.

6. Zeng Peiyan also says, "A major subject for study is what road to industrialization is to be taken. We must integrate industrialization and information, using information to stimulate industrialization and industrialization to stimulate the development of information, so as to pioneer a new road to industrialization featuring high-technology contents, low energy consumption, less pollution, and full use of human resources." Ibid., p. 81.

7. Full employment does not mean 100 percent employment. It means that the actual unemployment rate is lower than the natural unemployment rate, which is 5 percent, under China's specific conditions. Hu Angang, *Employment and Development: China's Unemployment Problems and Employment Strategy* (Shenyang: Liaoning People's Publishing House, 1998), pp. 34–35.

8. Hu Angang, "China's Human Development Trend and Long-Term Goal," *National Conditions Report*, no. 5 (2006).

9. Hu Angang and Zhang Ning, "Evolution of Human Development in Various Regions of China (1982–2003)," *China Studies Report*, no. 20 (2006).

10. *China Statistical Yearbook 2006*, appendix 2-6.

11. Drinking water failing to meet safety standards means that the drinking water contains high chlorine, arsenic, or bitter pollutants; is from areas infested with snail fever; contains microorganisms; or is locally scarce. Ma Kai, *Reader for Understanding the Eleventh Five-Year Plan* (Beijing: People's Publishing House, 2006), p. 515.

12. The average proportion spent on health among countries with midlevel HDI in 2000 is 6.9 percent of GDP. UN Development Program, "A Compact among Nations to End Human Poverty," *Human Development Report 2003*.

13. China's Tenth Five-Year Plan set fiscal expenditure in education at 4 percent of GDP, but it dropped due to an adjustment in the 2004 GDP base figure. By 2005 it was 2.82 percent of GDP. *China Education News*, June 6, 2007. The 2000 world elementary education level is the general national average. UN Development Program, "A Compact among Nations to End Human Poverty," and "Average of Countries with a Good Tertiary Education in 2000," in *Human Development Report 2003*.

14. Compulsory education in 2009 exceeded 99 percent, and gross enrollment rate in junior secondary schools was 99 percent. *China Statistical Abstract 2010*, p. 174. Illiteracy

among young and middle-aged people was 4 percent in 2005. Ma, *Reader for Understanding the Eleventh Five-Year Plan*, p. 515.

15. The Tenth Five-Year Plan set the goal of bringing the gross enrollment rate at the senior secondary level up to 80 percent by 2010 and the number of students to 45 million. Ma, *Reader for Understanding the Eleventh Five-Year Plan*, p. 515. Senior secondary schools include regular senior secondary, vocational secondary, adult secondary, regular professional, adult profession, and technical schools. *China Statistical Abstract 2006*, p. 189.

16. The Tenth Five-Year Plan set the goal of bringing the enrollment of intermediate vocational schools up to 8 million by 2010. During the eleventh plan there will be 25 million graduates from intermediate vocational schools. Ma, *Reader for Understanding the Eleventh Five-Year Plan*, p. 515.

17. The number of students in tertiary schools in 2005 was about 24 million, with the gross enrollment rate reaching 21 percent. *China Statistical Abstract 2006*, p. 189. The number is expected to reach 40 million by 2020.

18. China is already in the front ranks of the world in number of scientific and technical personnel, boasting 32 million technical personnel and 1 million R&D personnel. *Renmin ribao* [People's Daily], October 20, 2006. By the end of 2004 there were about 60 million personnel (including party and government personnel, business management personnel, and specialized technical personnel). Ma, *Reader for Understanding the Eleventh Five-Year Plan*, p. 515.

19. State Council, Outlined Program for the Medium- and Long-Term Development of S&T, 2006–2020. China ranked fifth in the world in terms of the number of papers in 2005. During the Tenth Five-Year Plan the number of patent applications reached 1.59 million, averaging an annual growth of 22.8 percent. Ma, *Reader for Understanding the Eleventh Five-Year Plan*, p. 515.

20. China's high-tech product exports in 2006 were valued at $281.5 billion, 29 percent of total merchandise export, or 1.3 times the high-tech exports of the United States, which was $216 billion. China has already become the number-one exporter of high-tech products.

21. The amount of water used by agriculture in 2000 was 68.8 percent of the total. It dropped to 63.6 percent by 2005. "Water Supply," in *China Statistical Yearbook 2006*.

22. Reserves in 2000 accounted for 12.8 percent of total land area. Reserves rose to 15 percent in 2005. *China Statistical Abstract 2006*, p. 203. The plan for 2010 is 16 percent. Ma, *Reader for Understanding the Eleventh Five-Year Plan*, p. 515.

23. Ming Wang, "Natural Disaster Insurance in China: Practice and Lessons Learned," paper prepared for presentation at Beijing Normal University, 2010.

24. According to Fortune 500 for 2006, there were 23 Chinese enterprises, including 19 on the mainland, 1 in Hong Kong, and 3 in Taiwan. The 19 mainland enterprises had a combined business volume of $594 billion, 3.13 percent of the total of the top 500 enterprises ($18,929.4 billion). Their combined profits were $40.6 billion, 3.34 percent of the total of the top 500 ($1,214.9 billion). Total profitability was 6.8 percent, higher than the 6.4 percent average of the top 500.

25. Hu Angang and Wang Yahua, *National Conditions and Development* (Beijing: Tsinghua University Press, 2005), p. 17.

26. Resolutions of the CCP Central Committee Concerning Major Issues about Building a Socialist Harmonious Society, adopted at the sixth plenary session of the Sixteenth CCP Central Committee, October 11, 2006.

27. State Council, Outlined Program for the Eleventh Five-Year National Economic and Social Development of the People's Republic of China, adopted at the fourth session of the Tenth National People's Congress, March 14, 2006.

28. *China Statistical Abstract 2009.*

29. Hu and Wang, *National Conditions and Development.*

30. *China Statistical Abstract 2009.*

31. Ibid., p. 176.

32. *China Statistical Abstract 2010,* pp. 24–25.

33. One indicator of population development is total fertility rate. For population sustainability it should be around 2.0; if it is lower than 1.8 or higher than 2.1, population development may not be sustainable.

FURTHER READING:
THE WRITINGS OF HU ANGANG, 1989–2011

2011

Guoqing Yanjiu yu Jiaoshu Yuren [Hu Angang's Research on China Studies and His Teaching and Training of the New Generation]. Tsinghua University Press.

2010

Yu Shijie Duihua [Dialogue with the World]. Shanghai: Orient Press.

Zhongguo: Zouxiang 2015 [China: Toward 2015], coauthored with Yan Yilong. Hangzhou: Zhejiang People's Press.

2009

Fanrong Wending Lun: Guojia Heyi Fuqiang Hexie [Theories on Prosperity and Stability: Why Countries Are Rich and Harmonious], coauthored with Hu Lianhe. Beijing: Encyclopedia of China Publishing House.

Zhongguo Jiaotong Geming: Kuayueshi Fazhan Zhilu [China's Communication Revolution: Leapfrogging Development], co-edited with Shi Baolin. Beijing: China Communications Press.

Zhongguo Yingdui Quanqiu Qihou Bianhua [China Combats Global Climate Change], coauthored with Guan Qingyou. Tsinghua University Press.

2008

Mao Zedong yu Wenge [Mao Zedong and the Culture Revolution]. Hong Kong: Strong Wind Press.

Zhongguo: Minsheng yu Fazhan [China: People's Livelihood and Development]. Beijing: China Economic Publishing House.

Zhongguo Zhengzhi Jingji Shilun (1949–1976) [Historical Review of Chinese Politics and Economics (1949–1976)]. Tsinghua University Press.

2007

Guoqing Baogao: Jingji Daguo Zhongguo de Keti [China Studies Report: Issues in China as an Economic Power]. Tokyo: Iwanami Shoten.

Hexie Shehui Goujian: Ouzhou de Jingyan yu Zhongguo de Tansuo [Constructing a Harmonious Society: European Experiences and China's Exploration], coedited with Zhou Jianming and Wang Shaoguang. Tsinghua University Press.

Zhongguo Jueqi Zhilu [Roadmap of China's Rise]. Peking University Press.

2020 Zhongguo: Quanmian Jianshe Xiaokang Shehui [China in 2020: Building a Well-Off Society]. Tsinghua University Press.

2006

Achievement Evaluation of IFI Assistance Loans to China (1981–2002), coauthored with Hu Guangyu. Hong Kong: Springer.

Economic and Social Transformation in China: Challenges and Opportunities. London: Routledge.

Zhongguo: Zai Shang Xin Taijie [China: Climbing One More Step]. Hangzhou: Zhejiang People's Press.

2005

Guoqing yu Fazhan: Zhongguo Wuda Ziben Dongtai Bianhua yu Changyuan Fazhan Zhanlue [National Conditions and Development: Dynamics of Five Capitals and Long-Term Development Strategy in China], coauthored with Wang Yahua. Tsinghua University Press.

Yuanzhu yu Fazhan: Guoji Jinrong Zuzhi dui Zhongguo Daikuan Jixiao Pingjia (1981–2002) [ODAs and Development: Performance Evaluation on Loans from IFIs to China (1981–2002)], coauthored with Hu Guangyu. Tsinghua University Press.

Zhongguo: Dongya Yitihua Xin Zhanlue [China: New Strategy for East Asia's Integration], coedited with Men Honghua. Hangzhou: Zhejiang People's Press.

Zhuanxing yu Wending: Zhongguo Ruhe Changzhi Jiuan [Transition and Stability: How China Achieves Long-Term Stability], coauthored with Hu Lianhe. Beijing: People's Publishing House.

2004

Zhongguo: Xin Fazhanguan [China: New Conceptions of Development]. Hangzhou: Zhejiang People's Press.

2003

Toushi SARS: Jiankang yu Fazhan [Perspective on SARS: Health and Development], ed. Tsinghua University Press.

Jiedu Meiguo Da Zhanlue [Understanding the U.S. Grand Strategy], coedited with Men Honghua. Hangzhou: Zhejiang People's Press.

Zhongguo Da Zhanlue [China's Grand Strategy], ed. Hangzhou: Zhejiang People's Press.

Zhongguo Dierdai Gaige Silu: Yi Zhifu Jianshe Wei Zhongxin [Reform Ideas of China's Second-Generation Leaders: Centered on Institutional Construction], coedited with Wang Shaoguang and Zhou Jianming. Tsinghua University Press.

Zhongguo Ruhe Ganchao Meiguo: Hu Wen Tizhi de Zhanlue [How China Will Overtake the U.S.: Hu and Wen's Strategy]. Tokyo: PHP Institute.

2002

Kuoda Jiuye yu Tiaozhan Shiye: Zhongguo Jiuye Zhengce Pinggu (1949–2001) [Increasing Employment and Meeting the Challenge of Unemployment: Evaluation of China's Employment Policies (1949–2001)], coauthored with Cheng Yonghong. Beijing: China Labor and Social Security Publishing House.

Quanqiuhua Tiaozhan Zhongguo [Globalization Challenges China], ed. Peking University Press.

Yingxiang Juece de Guoqing Baogao [China Studies Report: Influencing Decisions], ed. Tsinghua University Press.

Zhongguo Zhanlue Gouxiang [Ideas on China's Strategy], ed. Hangzhou: Zhejiang People's Press.

2001

Diqu yu Fazhan: Xibu Kaifa Xin Zhanlue [Regions and Development: New Strategies for the Development of the Western Regions], coedited with Zou Ping. Beijing: China Plan Press.

Nengyuan yu Fazhan: Quanqiuhua Tiaojian Xia de Nengyuan yu Huanjing Zhengce [Energy and Development: Energy and Environment Policy under Globalization], coedited with Lu Yonglong. Beijing: China Plan Press.

Zhishi yu Fazhan: 21 Shiji Xin Zhuigan Zhanlue [Knowledge and Development: New Catch-Up Strategies in the 21st Century], ed. Peking University Press.

Zhongguo: Tiaozhan Fubai [China: Challenging Corruption], ed. Hangzhou: Zhejiang People's Press.

Zhongguo Tiaozhan Tonghuo Jinsuo [China Challenges Deflation], coedited with Wu Qungang, Shen Bingxi, and others. Beijing: China Plan Press.

2000

Daguo Zhanlue [Strategies of a Great Power], coauthored with Yang Fan. Shenyang: Liaoning People's Press.

Shehui yu Fazhan: Zhongguo Shehui Fazhan Diqu Chaju Yanjiu [Society and Development: Study of the Regional Disparities in Social Development in China], coauthored with Zou Ping. Hangzhou: Zhejiang People's Press.

Zhongguo Zouxiang [China's Trend], ed. Hangzhou: Zhejiang People's Press.

Zhengfu yu Shichang [Government and Market], coedited with Wang Shaoguang. Beijing: China Plan Press.

1999

The Chinese Economy in Crisis: State Capacity and Tax Reform, coauthored with Wang Shaoguang. New York: M. E. Sharp.

The Political Economy of Uneven Development: The Case of China, coauthored with Wang Shaoguang. New York: M. E. Sharp.

Zhongguo: Bu Pingheng Fazhan de Zhengzhi Jingjixue [The Political Economy of Uneven Development: The Case of China], coauthored with Wang Shaoguang. Beijing: China Plan Press.

Zhongguo Fazhan Qianjing [China's Development Prospects]. Hangzhou: Zhejiang People's Press.

Zhongguo Nongye Chanyehua Lilun yu Shijian [Theory and Practice of Agriculture Industrialization in China], ed. Shenyang: Liaoning People's Press.

1998

Jiuye yu Fazhan: Zhongguo Shiye Wenti yu Jiuye Tiaozhan [Employment and Development: China's Unemployment Problems and Employment Strategies]. Shenyang: Liaoning People's Press.

Sikao Zhongguo: Tiaozhan Zhongguo de Jiu Da Wenti [Thinking about China: Nine Big Problems Challenging China], coauthored with Yang Yongfu. Shenyang: Liaoning People's Press.

1997

Zhongguo Ziran Zaihai yu Jingji Fazhan [Natural Disasters and Economic Development in China], coauthored with Lu Zhongchen, Sha Wanying, and others. Wuhan: Hubei Science and Technology Press.

1995

Hu Angang Ji: Zhongguo Zouxiang Ershiyi Shiji de Shida Guanxi [Selected Works of Hu Angang: Ten Significant Relations toward the 21st Century]. Harbin: Heilongjiang Education Press.

Tiaozhan Zhongguo: Deng hou Zhongnanhai de Jiyu yu Tiaozhan [Challenging China: Opportunities and Choices of Zhongnanhai after Deng]. Taipei: Journalist Co.

Zhongguo Diqu Chaju Baogao [Report on China's Regional Disparity], coauthored with Wang Shaoguang and Kang Xiaoguang, Shenyang: Liaoning People's Press.

Zhongguo Xiayibu [China's Next Step]. Chengdu: Sichuan People's Press.

1994

Zhongguo Jingji Bodong Baogao [Report on China's Economic Fluctuation]. Shenyang: Liaoning People's Press.

1993

Zhongguo Guojia Nengli Baogao [Report on China's State Capacity], coauthored with Wang Shaoguang. Shenyang: Liaoning People's Press.

1992

Survival and Development: A Study of China's Long-Term Development, coauthored with Zhou Lisan. Beijing: China Science Press, 1992.

1991

China's Population Development, coauthored with Zou Ping. Beijing: China Science and Technology Press.

Zhongguo Gongyehua Wenti Chutan [Tentative Exploration into China's Industrialization], coauthored with Guo Qing. Beijing: China Science and Technology Press.

Zhongguo: Zouxiang Ershiyi Shiji [China: Toward the 21st Century]. Beijing: China Environmental Science Press.

1990

Guoqing yu Juece [National Conditions and Decisionmaking], coauthored with Wang Yi and Zhao Tao. Beijing Publishing House.

Renlei, Fazhan, Qianjing, Jueze [Human Development, Prospect, Choice], coauthored with Zou Ping. Beijing: Academic Books.

1989

Renkou yu Fazhan: Zhongguo Renkou Jingji Wenti de Xitong Yanjiu [Population and Development: A Systematic Study on Population and Economic Problems in China]. Hangzhou: Zhejiang People's Press.

Shengcun yu Fazhan [Survival and Development], coauthored with Wang Yi. Beijing: Science Press.

INDEX

Acheson, Dean, 49
Acquired immunodeficiency syndrome (AIDS), 65, 74, 76, 146. *See also* Health and public health issues—China
Africa, 53
Agricultural issues—China: agricultural production and output, 10, 38, 40, 42, 43t, 50, 125–27; climate change, 128–29; history of, 26; hybrid rice, 118; transition of labor to the nonagricultural sector, 36–37
Agricultural issues—U.S., 40, 42, 43t
AIDS. *See* Acquired immunodeficiency syndrome
Alcohol and alcohol consumption, 73
Asia, 53, 112, 123
Asian Development Bank, 5
Asian Tigers (Hong Kong, Singapore, South Korea, Taiwan), 24, 25. *See also individual countries*
Australia, 87, 114, 130

Bairoch, Paul, 9
Bangladesh, 78

Beijing (China), 34, 78, 79
Beijing consensus, 17
Bo Yibo, 33
Brazil, 105
BRIC countries (Brazil, Russia, India, China), 5–6. *See also individual countries*
Britain. *See* United Kingdom
"Build China a Powerful Modern Socialist Country" (Mao; *1964*), 108

Canada, 114
Capital (human, natural, and physical)—China: educational factors in, 83, 85; future goals for, 39, 57, 119–20; growth of, 27, 28, 30, 35, 36, 69, 88–89, 90; in science and technology, 96
Capital (human, natural, and physical)—general: calculation of, 82; definition of, 27, 82; externalities and spillover effects of, 82; welfare and, 28
Carnegie Endowment for International Peace. *See* Keidel, Albert
CCP. *See* Chinese Communist Party

United Nations Trade and Development Commission, 112

United States (U.S.): agricultural output, 10; culture of, 17; demographic issues, 53, 65, 68; educational issues of, 36, 85, 87, 89, 147; environmental issues of, 129, 130, 131t, 135, 150; immigrant population in, 25; innovation in, 25; labor force of, 23, 96; national power of, 27; population issues of, 25, 53; PRC and, 14, 17, 20, 39–46; public health in, 68; research and development in, 96; social problems of, 16; urbanization in, 58–60, 62. *See also* Economic issues—U.S.

United States—Science and technology: information and communication technology (ICT), 113; Internet users in, 107; patent applications by, 102, 103, 104; personal computer users in, 106; published science and technology papers, 100–01, 114; research and development expenditures in, 105; as an S&T power, 99, 108, 109, 110, 148

Urban areas—China: causes of, 60–61; challenges, opportunities and development of, 61–64; climate change and, 122; death rates in, 67; economic factors of, 57, 58–59, 60, 63, 64; family planning policies and, 56; during the first baby boom, 50; future acceleration of urbanization, 63; health and health care in, 71–72, 73, 75; infrastructure in, 62; planned economy period and, 58–59; population factors in, 57, 59–62, 63–64; reform and opening and, 59–61; size of, 63; slums in, 63; social management and, 61, 63; standards of living in, 37–38; strategies of, 63, 64; urbanization, 8, 49, 57–64, 144

Urban areas—general, 64

USSR (Union of Soviet Socialist Republics). *See* Soviet Union

Wang Feng, 53
Wang Xiaolu, 35, 90
WHO. *See* World Health Organization
World Bank, 2–3, 7, 34, 35, 42, 62, 99, 114
World Energy Outlook 2007: China and India Insights (IEA), 129
World Health Organization (WHO), 73, 74
World Intellectual Property Organization, 102, 103
World Trade Organization (WTO), 10
World War II, 25
WTO. *See* World Trade Organization

Yuan Longping, 118

Zeng Yi, 56
Zhang Honglin, 60–61
Zhao Ziyang, 51
Zhou Enlai, 50, 109
Zhou Xiaochuan, 35